T0196108

MACAT

An Analysis of

James E. Lovelock's

Gaia:

A New Look at Life on Earth

Mohammad Shamsudduha

ROUTLEDGE

Published by Macat International Ltd
24:13 Coda Centre, 189 Munster Road, London SW6 6AW.

Distributed exclusively by Routledge
2 Park Square, Milton Park, Abingdon, Oxon OX14 4RN
711 Third Avenue, New York, NY 10017, USA

Routledge is an imprint of the Taylor & Francis Group, an informa business

www.macat.com
info@macat.com

Cataloguing in Publication Data
A catalogue record for this book is available from the British Library.
Library of Congress Cataloguing-in-Publication Data is available upon request.
Cover illustration: Etienne Gilfillan

ISBN 978-1-912302-38-3 (hardback)
ISBN 978-1-912128-09-9 (paperback)
ISBN 978-1-912281-26-8 (e-book)

Notice
The information in this book is designed to orientate readers of the work under analysis,
to elucidate and contextualise its key ideas and themes, and to aid in the development
of critical thinking skills. It is not meant to be used, nor should it be used, as a
substitute for original thinking or in place of original writing or research. References and
notes are provided for informational purposes and their presence does not constitute
endorsement of the information or opinions therein. This book is presented solely for
educational purposes. It is sold on the understanding that the publisher is not engaged
to provide any scholarly advice. The publisher has made every effort to ensure that
this book is accurate and up-to-date, but makes no warranties or representations with
regard to the completeness or reliability of the information it contains. The information
and the opinions provided herein are not guaranteed or warranted to produce particular
results and may not be suitable for students of every ability. The publisher shall not be
liable for any loss, damage or disruption arising from any errors or omissions, or from
the use of this book, including, but not limited to, special, incidental, consequential or
other damages caused, or alleged to have been caused, directly or indirectly, by the
information contained within.

CONTENTS

THE MACAT LIBRARY

The Macat Library is a series of unique academic explorations of seminal works in the humanities and social sciences – books and papers that have had a significant and widely recognised impact on their disciplines. It has been created to serve as much more than just a summary of what lies between the covers of a great book. It illuminates and explores the influences on, ideas of, and impact of that book. Our goal is to offer a learning resource that encourages critical thinking and fosters a better, deeper understanding of important ideas.

Each publication is divided into three Sections: Influences, Ideas, and Impact. Each Section has four Modules. These explore every important facet of the work, and the responses to it.

This Section-Module structure makes a Macat Library book easy to use, but it has another important feature. Because each Macat book is written to the same format, it is possible (and encouraged!) to cross-reference multiple Macat books along the same lines of inquiry or research. This allows the reader to open up interesting interdisciplinary pathways.

To further aid your reading, lists of glossary terms and people mentioned are included at the end of this book (these are indicated by an asterisk [*] throughout) – as well as a list of works cited.

Macat has worked with the University of Cambridge to identify the elements of critical thinking and understand the ways in which six different skills combine to enable effective thinking.
Three allow us to fully understand a problem; three more give us the tools to solve it. Together, these six skills make up the **PACIER** model of critical thinking. They are:

ANALYSIS – understanding how an argument is built
EVALUATION – exploring the strengths and weaknesses of an argument
INTERPRETATION – understanding issues of meaning

CREATIVE THINKING – coming up with new ideas and fresh connections
PROBLEM-SOLVING – producing strong solutions
REASONING – creating strong arguments

To find out more, visit **WWW.MACAT.COM.**

CRITICAL THINKING AND *GAIA: A NEW LOOK AT LIFE ON EARTH*

Primary critical thinking skill: INTERPRETATION
Secondary critical thinking skill: REASONING

Gaia: A New Look At Life on Earth may continue to divide opinion, but nobody can deny that the book offers a powerful insight into the creative thinking of its author, James E. Lovelock.

Published in 1979, Gaia offered a radically new hypothesis: the Earth, Lovelock argued, is a living entity. Together, the planet and all its separate living organisms form a single self-regulating body, sustaining life and helping it evolve through time. Lovelock sees humans as no more special than other elements of the planet, railing against the once widely-held belief that the good of mankind is the only thing that matters.

Despite being seen as radical, and even idiotic on its publication, a version of Lovelock's viewpoint has found resonance in contemporary debates about the environment and climate, and has now broadly come to be accepted by modern thinkers. As man's effects on the climate become increasingly extreme, more and more elements of the Earth's self-regulation seem to be unveiled – forcing scientists to ask how far the planet might be able to go in order self-regulate effectively. Indeed, despite its far-fetched elements, Lovelock's Gaia thesis seems to ring more convincingly today than ever before; that it does is largely a result of the critical thinking skills that allowed Lovelock to produce novel explanations for existing evidence and, above all, to connect existing fragments of evidence together in new ways.

ABOUT THE AUTHOR OF THE ORIGINAL WORK

James Lovelock was born in England in 1919, to working-class parents with little education. Despite financial struggles early in life, | he earned several degrees. Lovelock then invented a number of scientific instruments, which gave him the financial means to do research outside the university system. While working with NASA (the National Aeronautics and Space Administration) he grew fascinated by how life and the environment interact. In 1979, his first book Gaia: A New Look at Life on Earth put forward the controversial idea that our planet is self-regulating. The debate about Lovelock's 'Gaia hypothesis' has simmered for decades. He continues to expand and defend his work, and in 2015 published his tenth book on Gaia, at the age of 95.

ABOUT THE AUTHORS OF THE ANALYSIS

Dr Mohammad Shamsudduha holds a PhD in hydrogeology from University College, London. He is currently a researcher at the University College, London, Institute for Risk and Disaster Reduction.

ABOUT MACAT

GREAT WORKS FOR CRITICAL THINKING

Macat is focused on making the ideas of the world's great thinkers accessible and comprehensible to everybody, everywhere, in ways that promote the development of enhanced critical thinking skills.

It works with leading academics from the world's top universities to produce new analyses that focus on the ideas and the impact of the most influential works ever written across a wide variety of academic disciplines. Each of the works that sit at the heart of its growing library is an enduring example of great thinking. But by setting them in context – and looking at the influences that shaped their authors, as well as the responses they provoked – Macat encourages readers to look at these classics and game-changers with fresh eyes. Readers learn to think, engage and challenge their ideas, rather than simply accepting them.

'Macat offers an amazing first-of-its-kind tool for interdisciplinary learning and research. Its focus on works that transformed their disciplines and its rigorous approach, drawing on the world's leading experts and educational institutions, opens up a world-class education to anyone.'

Andreas Schleicher
Director for Education and Skills, Organisation for Economic Co-operation and Development

'Macat is taking on some of the major challenges in university education ... They have drawn together a strong team of active academics who are producing teaching materials that are novel in the breadth of their approach.'

Prof Lord Broers,
former Vice-Chancellor of the University of Cambridge

'The Macat vision is exceptionally exciting. It focuses upon new modes of learning which analyse and explain seminal texts which have profoundly influenced world thinking and so social and economic development. It promotes the kind of critical thinking which is essential for any society and economy. This is the learning of the future.'

Rt Hon Charles Clarke, former UK Secretary of State for Education

'The Macat analyses provide immediate access to the critical conversation surrounding the books that have shaped their respective discipline, which will make them an invaluable resource to all of those, students and teachers, working in the field.'

Professor William Tronzo, University of California at San Diego

WAYS IN TO THE TEXT

KEY POINTS

- James E. Lovelock, one of the world's foremost independent scientists and an outstanding popular science book writer, was born in 1919 into a working-class* family in the southeast of England.

- According to Lovelock's *Gaia: A New Look at Life on Earth*, the planet's living beings interact with the atmosphere, oceans, and rocks to form a self-regulating,* stable biosphere* where life can flourish ("biosphere" here refers to those parts of a planet occupied by living beings).

- Lovelock's controversial Gaia hypothesis*—according to which the Earth is a living superorganism* (an entity, like a colony, made up of many distinct individuals or organisms)—has led to the development of an entirely new academic discipline: Earth system science.*

Who Was *James E. Lovelock?*

James E. Lovelock, the author of *Gaia: A New Look at Life on Earth* (1979), was born in 1919 in the southeast of England into a working-class family.[1] His father Thomas A. Lovelock* was an art dealer; his mother Nellie A. Elizabeth* worked in city administration as a personal secretary;[2] neither had any serious formal education. Nellie had to leave school at the age of 13 to earn her living while Thomas had never been to school as a child, and could not read or write before he attended a technical college later in his life.[3] Perhaps on account of their history, Lovelock's parents valued education highly and always encouraged their son to go to school—but financial hardship in the family meant he could not afford to go to college when he came of age. Instead, he took a training position with a firm of chemical consultants while attending evening classes.

Lovelock eventually went to the University of Manchester[4] with a scholarship and graduated in chemistry* in 1941. In 1948, he received his doctor's degree in medicine. In 1959, he received a doctorate in biophysics (a discipline in which the science of physics is applied to the study of organisms) from the University of London.[5]

Working for most of his life as an independent scientist, Lovelock has been described by the British *Guardian* newspaper as a "world-famous author and speaker."[6] He owes this fame to his radical Gaia hypothesis: the proposition that the Earth can be considered an entity composed of many distinct individuals or organisms—a living superorganism comparable to a colony of bees or ants. Referring to Charles Darwin,* the nineteenth-century founder of evolutionary* theory, the English philosopher Mary Midgley* has written of Lovelock, "Though fanatically accurate over details, he never isolates those details from a wider, more demanding vision of their background. He thinks big. Preferring, as Darwin did, to work outside the tramlines of an institution, he has supported himself since 1963 through inventions and consultancies."[7]

What Does *Gaia: A New Look at Life on Earth* Say?

Having discussed the idea in academic papers from 1969 onward,[8] Lovelock began to write *Gaia: A New Look at Life on Earth* in 1974, while living in an area of natural beauty in western Ireland. The book, his first, is written in simple, nontechnical language to convey his ideas to a nonacademic audience; in it, he introduced the general public to the rather controversial idea of our planet as a self-regulating and living entity, which he called "Gaia" after the ancient Greek* goddess of the Earth.

For Lovelock, humans are not a special feature of the planet, being part of a broader community of living creatures. Referencing the political and philosophical position of humanism,* emphasizing the importance of human affairs, Lovelock writes, "I began more and more to see things though her [Gaia's] eyes and slowly dropped off, like an old coat, my loyalty to the humanist Christian belief in the good of mankind as the only thing that matters."[9]

According to Lovelock's hypothesis,* the living Earth is a superorganism—a living thing made up of many other living things all interacting both mutually and with the air, the oceans, and the planet's surface rocks. This system of mutual interaction has the effect of making Earth a fit and comfortable place to live. Suggesting for the first time that living organisms control their nonliving environment, or surroundings, the idea has proven to be controversial.

Lovelock laid out his thoughts as a story in a series of interlinked chapters. Because of his simple writing style, shortage of evidence, and his use of myth and poetry, the book received very harsh reviews[10] within the scientific community when it was first published in 1979. Scientists from several disciplines, among them biology* and geology,* expressed serious reservations about the idea. (Biology is the study of living organisms; geology is the study of the formation and structure of the planet's physical material, such as rock.)

The essence of Lovelock's Gaia hypothesis is that the entire surface of the Earth, including all living organisms, is a self-regulating body. It makes changes to itself as needed to maintain a balance between its physical, chemical, and biological environments, thereby sustaining life and helping it evolve through time. Earth's lasting tendency to maintain a stable condition for its living creatures through self-regulation is known as "homeostasis."* This homeostatic feature of the planet Earth is the basis for the "Gaia hypothesis"—a strongly debated topic within scientific and philosophical circles.[11]

Lovelock's Gaia hypothesis continues to be relevant in the scientific arena because of its relationship to the ongoing discussion and controversy about climate change*—the large-scale, long-term shift in the planet's weather patterns or average temperatures that, according to the consensus, human action has provoked. A search for the phrase "Gaia hypothesis" on the scientific database ScienceDirect* returns 373 scientific articles and 146 books. Additionally, a search for the phrase "Gaia hypothesis" on Google search engine returns over 400,000 pages on the Internet.

Why Does *Gaia: A New Look at Life on Earth* Matter?

Lovelock's Gaia hypothesis has been influencing academics, scientists, politicians, and the general public since the 1980s. While the idea of a self-regulating, living Earth was initially rejected by many scientists, so was the German scientist Alfred Wegener's* theory that the Earth's continents drift and move position over the planet's surface ("continental drift")* and the English naturalist Charles Darwin's theory of evolution by means of natural selection, explaining how environment, adaptation, and inherited characteristics lead to the formation of new species. Both are accepted as fact today.[12]

Initially, Lovelock's hypothesis was severely criticized by geologists, evolutionary biologists* (those studying the Earth's living things in the light of evolutionary theory), and planetary scientists* (those engaged in the scientific study of planets and moons). For example, the evolutionary biologist Richard Dawkins* argued that organisms could not act as a whole to control their environment.[13]

Despite the widespread criticism, many scientists thought the idea was worth discussing further, and Lovelock's Gaia hypothesis started to receive more favorable attention in the late 1980s.[14] In 1985, the University of Massachusetts hosted the first public symposium on the Gaia hypothesis, titled "Is the Earth a Living Organism?"; in 1988, the American Geophysical Union* (AGU)'s First Chapman Conference on the Gaia Hypothesis was held in San Diego, California.[15] "Change is in the air," Lovelock wrote in 1994 when a scientific meeting on "The Self-regulating Earth" was held in the English city of Oxford. In 2000, the AGU held a second conference in Valencia, Spain;[16] in 2006, an international conference on the Gaia hypothesis was held at George Mason University in Virginia.[17]

In response to the initial criticism, Lovelock and his principal collaborators, the US geoscientist* Lynn Margulis* and the British marine and atmospheric scientist Andrew Watson,* developed the Gaia hypothesis into a theory that might be scientifically tested ("geoscience" is the branch of science drawing on geology, physics, and chemistry to consider the defining attributes of the Earth). Although the terms "Gaia hypothesis" and "Gaia theory" are often used interchangeably, Lovelock used them with separate meanings. For him, the hypothesis became a theory only after there began to be scientific evidence to support it—in his *Ages of Gaia* (1995), he writes that "we can now begin to think of Gaia as a theory, something rather more than mere 'let's suppose' of a hypothesis."[18]

"To understand even the atmosphere, the simplest of the planetary compartments," Lovelock has pointed out, "it is not enough to be a geophysicist;* knowledge of chemistry and biology is also needed"[19] (geophysics is the study of the various gravitational, magnetic, electrical, and seismic phenomena such as earthquakes that define our planet). The core ideas of the Gaia hypothesis were so broad and varied that it was not apparent immediately which major scientific discipline could accommodate them. So Lovelock proposed that such a broad topic had to be discussed under a new interdisciplinary branch of science (that is, a branch of science drawing the aims and methods of different fields of scientific inquiry).

The Gaia hypothesis ultimately gave birth to a new academic discipline known as geophysiology*—a field, also known as Earth system science, that studies the interaction of the Earth's living things and the Earth itself. Many top-ranked academic institutions around the world, including Stanford University in California, have established Earth system science departments to study the planet's oceans, lands, and atmosphere as an integrated system.

NOTES

1   Ian Irvine, "James Lovelock: The Green Man," The *Independent*, December 3, 2005, accessed October 10, 2013, http://www.independent.co.uk/news/people/profiles/james-lovelock-the-green-man-517953.html.

2   James E. Lovelock, *Homage to Gaia: The Life of an Independent Scientist*, rev. ed. (London: Souvenir Press Ltd., 2014), 1.

3   James E. Lovelock, *The Vanishing Face of Gaia: A Final Warning* (London: Penguin Books, 2010), 206.

4   Robin McKie, "Gaia's Warrior," *Green Lifestyle Magazine*, July/August 2007, 60–62.

5   "Curriculum Vitae," James Lovelock's official website, accessed December 29, 2013, http://www.jameslovelock.org/page2.html.

6   Peter Forbes, "Jim'll Fix it," The *Guardian*, February 21, 2009, accessed December 23, 2013, http://www.theguardian.com/culture/2009/feb/21/james-lovelock-gaia-book-review.

7   Mary Midgley, "Great Thinkers—James Lovelock," *New Statesman*, 14 July 2003.

8   James E. Lovelock and C. E. Giffin, "Planetary Atmospheres: Compositional and Other Changes Associated with the Presence of Life," *Advances in the Astronautical Sciences*, 25 (1969): 179–93.

9   See James E. Lovelock, "Preface," in *Gaia: A New Look at Life on Earth*, by James E. Lovelock, rev. ed. (Oxford: Oxford University Press, 2000), ix.

10 W. Ford Doolittle, "Is Nature Really Motherly?," *CoEvolution Quarterly* 29 (1981): 58–65.

11 Richard R. Wallace and Bryan G. Norton, "Policy implications of Gaian Theory," *Ecological Economics* 6 (1992): 103.

12 Martin Ogle, "The Gaia Theory: Scientific Model and Metaphor for the 21st Century," *Revista Umbral (Threshold Magazine)* 1 (2009): 99–106.

13 See Richard Dawkins, *The Extended Phenotype: The Gene as the Unit of Selection* (Oxford: Oxford University Press, 1982), 234–36.

14 Lovelock, *Gaia: A New Look at Life on Earth*, xii.

15 Eric G. Kauffman, "The Gaia Controversy: AGU's Chapman Conference," *Eos, Transactions of the American Geophysical Union* 69 (1989), 763–64.

16 Brent F. Bauman, "The Feasibility of a Testable Gaia Hypothesis" (BSc thesis, James Madison University, 1998).

17 "Gaia Theory Conference at George Mason University," Arlington County, accessed December 27, 2013, http://www.gaiatheory.org/2006-conference/.

18 James E. Lovelock, *The Ages of Gaia: A Biography of Our Living Earth*, rev. ed. (Oxford: Oxford University Press, 1995), 44.

19 James E. Lovelock, "Geophysiology, the Science of Gaia," *Reviews of Geophysics* 27 (1989): 222.

SECTION 1
INFLUENCES

MODULE 1
THE AUTHOR AND THE HISTORICAL CONTEXT

KEY POINTS

- *Gaia: A New Look at Life on Earth* explains the concept of a self-regulating* Earth, a concept that has remained an original and important contribution to scientific debate around sustaining life on our planet; "self-regulating" describes the interactions of the Earth's organisms, air, water, and rocks in order to keep the planet fit and comfortable for life.

- Lovelock, an independent scientist through much of his career, says that the ability to think freely was critical to the development of the Gaia hypothesis,* according to which "the physical and chemical condition of the surface of the Earth, of the atmosphere, and of the oceans, has been and is actively made fit and comfortable by the presence of life itself."[1]

- Lovelock's short-term employment with NASA's* Jet Propulsion Laboratory in California on the unmanned Viking Mission to Mars* left him fascinated by the study of life on Earth and other planets (NASA is the organization responsible for the United States' civilian space program).

Why Read this Text?

James E. Lovelock's *Gaia: A New Look at Life on Earth* (1979) is one of the best-selling popular books on a contemporary scientific debate concerning life and planetary processes. In it, he describes a self-regulating, living Earth that he calls Gaia—after the ancient Greek* goddess of the Earth. Lovelock's radical Gaia hypothesis was fiercely

> **"** Humanity and science were offered a cornucopia of benefits from the accelerated inventions of the Second World War. Had we been less combative animals we could have used this new knowledge constructively. We could have made the observation of the Earth from space a priority, built satellites that viewed the land, the air, and the oceans, and seen the looming dangers of global warming in time; instead we made space missiles. **"**
>
> James E. Lovelock, *A Rough Ride to the Future*

criticized by evolutionary biologists,* chemists* (those with a specialist knowledge of the properties and reactions of matter), and planetary scientists* (those who study planets and moons). Even so, it has been proposed that the hypothesis has since been developed into a scientifically testable theory—the Gaia theory.*[2]

Over the last three decades, the Gaia hypothesis has been the subject of numerous scientific papers and books, and several international conferences.[3] Scientists and researchers from around the world recognized in the Amsterdam Declaration on Global Change* of 2001 that "the Earth system behaves as a single, self-regulating system comprised of physical, chemical, biological, and human components."[4] Although the Declaration does not mention Gaia, this is the key idea of the hypothesis.

The development by the Intergovernmental Panel on Climate Change (IPCC)* of Earth system models* to project future climate conditions has been a major achievement of the Gaia theory.[5] The IPCC is an intergovernmental body led by the United Nations;* Earth system models are used to evaluate regional and global climate under a number of different conditions by considering interactions between atmosphere, ocean, land, ice, and biosphere.*

Author's Life

James E. Lovelock was born on July 26, 1919 in Letchworth Garden City, England, into a working-class* family.[6] His father, Thomas A. Lovelock* had an interest in painting;[7] his mother, Nellie A. Elizabeth,* worked as a secretary. In his autobiography *Homage to Gaia*, Lovelock writes that "the bitterest blow for [my mother] came when she won a rare scholarship ... to a grammar school [a selective state school]. She could not take it because the family needed her earning power at thirteen to survive. Instead of an enlightened education ... she spent her days in a pickle factory sticking labels on the jars."[8] She was "full of working-class good intentions," he continues, "and she had an unquestioning belief in education. She was determined that I should go to a grammar school and as soon as possible. She had been denied the chance of a 'good education' and she did not intend that I should suffer from a lack of it."[9] While Lovelock did not particularly enjoy his school life, he was determined to become a scientist.

Lovelock started his professional career at the National Institute for Medical Research* in London. He worked there for nearly 20 years, with some career breaks. Never wanting to become a permanent employee of any educational or research organization, he became a fully independent scientist in the early 1960s; he has not been formally associated with any major university or research facility since then and has practiced science independently using the revenue earned from his inventions and publications. He writes that "one of the joys of independence is the extent to which the needs of different customers are shared in common: work done for one agency like NASA, often cross-fertilized the work I did for another, such as [the multinational oil and gas company] Shell."[10] Lovelock said this independence helped him tremendously in developing the radical Gaia hypothesis.

Author's Background

Lovelock's childhood, educational background, real-world experience, and many of his research collaborators immensely influenced his unusual way of thinking about life on Earth. From his childhood onward, he was particularly interested in nature and science. According to the *Green Lifestyle Magazine*,[11] his passion for science began with trips to science and natural history museums in London and with reading stories by the science fiction writers H. G. Wells,* the English author of *The War of the Worlds* (1898), and Jules Verne,* the French author of *From Earth to the Moon* (1865). Lovelock's own family greatly shaped his way of thinking and caring for nature. His greatest influence was his beloved father who expressed a great love for and care of nature throughout his life.[12] Lovelock describes in the book that his father used to say that every living creature on the surface of the Earth serves a certain purpose, and together they form a greater ecosystem* (a biological system made up of all the organisms found in a specific place) in which the physical environment interacts with living things.

Lovelock's quest for Gaia began in the early 1960s when he was working at the National Aeronautics and Space Administration (NASA) Jet Propulsion Laboratory on the Viking Mission* for exploring life on Mars. The unmanned space exploration was based on the theory that any evidence discovered for life on Mars would be similar to that for life on Earth. Lovelock proposed that a simple examination of the Martian atmosphere could tell if there was any life there. Soon, he became especially interested in and engaged with the subject of what life, precisely, is, and was motivated to conduct further research on Earth's atmosphere, the composition of oceans, and the role of living organisms* in the framework of planetary processes.

Lovelock's idea of a self-regulating Earth came to him in a time when there was no public concern about global climate change* (long-term change in the planet's weather patterns and local and global temperatures) or biodiversity* (the wide variety of life on the planet). His Gaia hypothesis offered an explanation for how greenhouse gases* (gases that trap the Sun's energy), protected early life-forms on Earth by keeping the atmosphere warm and comfortable; "the dangers of habitat destruction and inflation of the air with greenhouse gases," he has written, "seemed remote and trivial in the 1970s and 1980s."[13]

NOTES

1   James E. Lovelock, Gaia: A New Look at Life on Earth rev. ed. (Oxford: Oxford University Press, 2000), 144.

2   James W. Kirchner, "The Gaia Hypothesis: Conjectures and Refutations," Climatic Change 58 (2003): 21.

3   "Gaia Hypothesis," Environment website, accessed December 23, 2013, http://www.environment.gen.tr/gaia/70-gaia-hypothesis.html.

4   James E. Lovelock, "The Living Earth," Nature 426 (2003): 769–70.

5   James E. Lovelock, A Rough Ride to the Future (London: Penguin Books, 2015), 94.

6   Ian Irvine, "James Lovelock: The Green Man," The Independent, December 3, 2005, accessed October 10, 2013, http://www.independent.co.uk/news/people/profiles/james-lovelock-the-green-man-517953.html.

7   James E. Lovelock, Homage to Gaia: The Life of an Independent Scientist, rev. ed. (London: Souvenir Press Ltd., 2014), 7–10.

8   Lovelock, Homage to Gaia, 7–37.

9   Lovelock, Homage to Gaia, 15–16.

10 Lovelock, Homage to Gaia, 282.

11 Robin McKie, "Gaia's Warrior," Green Lifestyle Magazine, July/August 2007, 60–62.

12 Lovelock, Homage to Gaia, 8.

13 See James E. Lovelock, "Preface," in Gaia: A New Look at Life on Earth, viii.

MODULE 2
ACADEMIC CONTEXT

KEY POINTS

- When Lovelock worked at the United States space agency NASA,* he was ridiculed for his ideas about testing for life on Mars, and the scientific community initially ignored his Gaia hypothesis* (according to which the surface of the Earth and the life it supports are a self-regulating* entity).

- Lovelock was unaware that earlier scientists had briefly discussed the idea of a self-regulating Earth when he began to develop the Gaia hypothesis ("self-regulating" here means that the mutual interactions of the Earth's organic and nonorganic features keep the planet fit and comfortable for life).

- With one of his students, Lovelock created a mathematical model of an environment he called Daisyworld* to test his ideas.

The Work In Its Context

James E. Lovelock's *Gaia: A New Look at Life on Earth* is an original contribution to the understanding of the atmosphere and the role of organisms* in influencing the Earth's climate. The late 1950s and the 1960s were a new era in space exploration.* In 1957, the Soviet Union* launched Sputnik,* the first ever artificial satellite* to orbit the Earth.[1] The United States had a long rivalry with the Soviet Union and, in 1961, President John F. Kennedy* began expanding the US space program through NASA. He set a target to land a man on the Moon and return him safely home by the end of the decade.[2]

NASA was also preparing to launch spacecraft to Mars to search for life. In those days, little research was conducted on life beyond

> **❝** I expected to discover somewhere in the scientific literature a comprehensive definition of life as a physical process, on which one could base the design of life-detection experiments, but I was surprised to find how little had been written about the nature of life itself. **❞**
>
> James E. Lovelock, *Gaia: A New Look at Life on Earth*

Earth; biologists* (those engaged in the scientific study of organisms) designed experiments to try to copy the conditions on Mars, but these experiments had to be conducted on Earth. Lovelock, employed at NASA, was skeptical about experiments conducted on Earth to detect life on Mars; as an alternative to sampling Mars's soil to detect any presence of life, he proposed sampling the planet's atmosphere. His idea was rejected and ridiculed at NASA.[3]

Similarly, he failed to stimulate interest in the scientific community when he first presented his Gaia hypothesis at a conference in the 1960s. The idea of a self-regulating Earth was complex and could not be placed in any single academic field. It was not until he published *Gaia* in 1979 that the academic community took interest in his hypothesis.

Overview of the Field

Although Lovelock's self-regulating Earth was not discussed in any formal academic discipline before he introduced the idea,[4] as he writes, "the idea that the Earth is alive in a limited sense is probably as old as humankind."[5] Several great scientists in the past looked at the planet Earth as a living body. In 1785, James Hutton,* a famous Scottish geologist,* described Earth as a self-regulating system. Hutton, who said Earth was like a living creature, compared the cycling of Earth's nutrients (food) between soil and plants, and the movement of water from oceans to land and back, with the circulation of blood in a human body.[6] In *The Ages of Gaia* (1995), Lovelock writes, "James Hutton is rightly remembered as a deeply influential figure in the field of geology but his idea of a living Earth was forgotten, or denied, in the intense reductionism* of the nineteenth century"[7] ("reductionism" occurs when we see something complex as merely the sum of its parts, without considering how those parts may interact with each other). In the early twentieth century the Russian scientist Vladimir I. Vernadsky* discussed the notion of a living Earth with an envelope of life and popularized the term "biosphere"* to describe the area of living matter.[8]

Lovelock, however, was unaware of these ideas when he formulated his hypothesis, which developed the idea of a self-regulating Earth more than anyone had previously done. Lovelock's Gaia hypothesis was considered by many scientists as contradictory to the deeply influential English naturalist Charles Darwin's* theory of evolution,* which suggested that organisms evolve but their nonliving environment does not. Later, however, Lovelock demonstrated that living organisms could interact with their nonliving environments to achieve the best living conditions for themselves.[9]

Academic Influences

Lovelock's childhood experiences and relationship with his parents, his education, his experience in the working world, and his collaborators among research scientists all had a major influence on the unusual perspective he developed in thinking about life on Earth. He was also influenced by several early scientists; in his autobiography, he wrote that he was greatly influenced by the English American zoologist* George E. Hutchinson's* work on the Earth's biochemistry* (a zoologist is engaged in the scientific study of animals; biochemistry is the study of the chemical processes of living things). Hutchinson, who studied the interaction of organisms and their environment, described the Earth as a self-regulating body from the viewpoint of chemical activity.[10] Lovelock also wrote that his friend Sidney Epton,* a chemist, helped stir the first real public interest in Gaia by coauthoring an article in the popular science journal *New Scientist* in 1975.[11]

The concept of a self-regulating Earth and the development of the Gaia hypothesis were greatly shaped and augmented by Lovelock's main collaborators—the geoscientist* Lynn Margulis,* the philosopher Dian Hitchcock* and Epton. Lovelock and Margulis started working together on the Gaia hypothesis in the early 1970s. Lovelock describes this partnership as a "most rewarding scientific collaboration" in his book that led to the publication of their first joint scientific paper on the topic.[12] Margulis was not only a research collaborator but a good friend who believed in Lovelock so much that she helped him secure funding for research on the Gaia hypothesis and for publication of his second book, *The Ages of Gaia* (1995).[13] Later on, Andrew Watson,* a British PhD candidate, worked with Lovelock on his research. Watson and Lovelock together developed a mathematical model of an imaginary Earthlike planet, "Daisyworld,"*[14] in an effort to demonstrate that living organisms can actually control the atmosphere and climate in which they live.

NOTES

1   James E. Lovelock, *The Ages of Gaia: A Biography of Our Living Earth*, rev. ed. (Oxford: Oxford University Press, 1995), 4.

2   "Space Program," John F. Kennedy Presidential Library and Museum, accessed January 8, 2016, http://www.jfklibrary.org/JFK/JFK-in-History/Space-Program.aspx.

3   James E. Lovelock, *Homage to Gaia: The Life of an Independent Scientist*, rev. ed. (London: Souvenir Press Ltd., 2014), 250.

4   Lovelock, *The Ages of Gaia*, 3–14.

5   Lovelock, *The Ages of Gaia*, 9.

6   Lovelock, *The Ages of Gaia*, 9.

7   Lovelock, *The Ages of Gaia*, 9.

8   Lovelock, *The Ages of Gaia*, 10.

9   Lovelock, *The Ages of Gaia*, 41–61.

10 Lovelock, *Homage to Gaia*, 263.

11 James E. Lovelock and Sidney Epton, "The Quest for Gaia," *New Scientist* 65, no. 935 (1975): 304–09.

12 James E. Lovelock and Lynn Margulis, "Atmospheric Homeostasis by and for the Biosphere: The Gaia Hypothesis," *Tellus* 26, nos. 1–2 (1974): 1–10.

13 Lovelock, *Homage to Gaia*, 369.

14 Andrew J. Watson and James E. Lovelock, "Biological Homeostasis of the Global Environment: The Parable of Daisyworld," *Tellus* 35B (1983): 286–89.

MODULE 3
THE PROBLEM

KEY POINTS

- In the mid-1960s when Lovelock conceived the idea of a self-regulating Earth,* researchers at the United States space agency NASA* and academics were primarily interested in detecting life on other planets by using well-established biological experiments that were only tested on Earth.

- Lovelock's Gaia hypothesis* holds that every organism* on Earth is tightly coupled with its environment, and that the sum of these interactions maintains a suitable living condition.

- Earlier scientists had suggested in passing that Earth is a self-regulating, living system, but Lovelock devoted much of his life to trying to prove it.

Core Question

James Lovelock's *Gaia: A New Look at Life on Earth* should be read in the light of the author's preoccupation with the questions "What is life?" and "How should it be recognized?"

These core questions, although simple in nature, were indeed original and critical for the development of Lovelock's ideas of the evolution of life and the Earth as a single living entity. He spent many years developing his radical and groundbreaking idea into the Gaia hypothesis that every living organism on Earth—ranging from a tiny, microscopic virus to the largest whale—can be regarded as tightly coupled with its environment as a single entity. These coupled entities are capable of maintaining a suitable living condition by controlling the Earth's atmosphere and the climate.

> **❝** … [T]hinking about life on Mars gave some of us a fresh standpoint from which to consider life on Earth and led us to formulate a new, or perhaps revive a very ancient, concept of the relationship between the Earth and its biosphere. **❞**
>
> James E. Lovelock, *Gaia: A New Look at Life on Earth*

Lovelock's formulation of the hypothesis was influenced by his time at NASA's Jet Propulsion Laboratory in the mid-1960s. There, he found himself fascinated by the images of Earth taken by the astronauts on the agency's manned Apollo* space missions, which, he writes, led him to look at Earth's surface from the top down rather than from the bottom up. The thought of a living planet came to him while working on experiments for detecting life on other planets. He thought there must be a planet-scale regulation system that has always kept the Earth fit and comfortable for life, while Earth's closest neighbors, Mars and Venus, were lifeless.[1] He challenged his fellow scientists at NASA on the usefulness of conducting experiments for detecting life on Mars from the soil, proposing as an alternative that experiments to determine the composition of Mars's atmosphere would indicate the presence or absence of life.[2]

The Participants

In the mid-1960s, when Lovelock conceived the idea of a living Earth, there was no ongoing debate about this topic. Space exploration was at its peak, fueled by the rivalry between the Soviet Union* and the United States—ideologically opposed superpowers, then vying for global military and cultural dominance. At NASA, biologists* and soil scientists were designing life-detecting experiments for spacecraft; these experiments, however, were conducted on the surface of Earth. One of the scientists was Vance Oyama,* a biochemist* who insisted that Martian soil should be collected and examined for the presence of life. At that time, the definition of "life" and whether it can control the living environment on Earth's surface were not discussed in any traditional academic discipline.[3]

Lovelock's concept of a living Earth, Gaia, is closely linked to the concept of life; indeed, to understand Gaia, one must understand life. When he started his in-depth research on life and its significance on Earth's environment, Lovelock found that several early scientists had suggested that the Earth was alive. In 1785, the Scottish geologist* James Hutton,* known as the father of geosciences,* proposed that Earth was a superorganism*—a living entity composed and defined through the mutual interactions of the organisms inhabiting it. In the late nineteenth century, the English biologist Thomas H. Huxley* proposed that the planet Earth was a living, self-regulating system, while the Russian geochemist* Vladimir I. Vernadsky* suggested that life had shaped the Earth as an active geological force.[4] None of these early scientists, however, looked any further than their initial thoughts. Lovelock, however, took a fresh approach to this idea and made it his life's work. He created the Gaia hypothesis that explains how the Earth maintains a condition that is always fit and comfortable for life. This stability in the midst of constant change is called "homeostasis."*

The Contemporary Debate

The idea of a living Earth, although conceived by previous scientists, was not accepted by mainstream scientists, and was unknown outside a few old scientific publications. The Ukrainian philosopher and independent scientist Yevgraf M. Korolenko,* for instance, declared the Earth to be a living organism near the end of the nineteenth century.[5] However, when Lovelock himself conceived the idea of a living Earth and started asking the same questions he was not aware of these old statements or publications. There was no ongoing debate and no publications directly useful to answering the central questions of the self-regulating Earth.

The subject matter was so broad that Lovelock needed advice and collaboration to develop his hypothesis into a theory that was capable of making testable scientific predictions.[6] While many scientists simply criticized his idea, some agreed to collaborate with him: Dian Hitchcock,* Sidney Epton,* and Lynn Margulis,* particularly, actively helped him develop the Gaia hypothesis.

NOTES

1   James E. Lovelock, *Gaia: A New Look at Life on Earth*, rev. ed. (Oxford: Oxford University Press, 2000), 1–29.

2   James E. Lovelock, *Homage to Gaia: The Life of an Independent Scientist*, rev. ed. (London: Souvenir Press Ltd., 2014), 242–43.

3   James E. Lovelock, *The Ages of Gaia: A Biography of Our Living Earth*, rev. ed. (Oxford: Oxford University Press, 1995), 15–20.

4   See Crispin Tickell, "Foreword," in *The Revenge of Gaia: Why the Earth is Fighting Back—and How We Can Still Save Humanity*, by James E. Lovelock (London: Penguin Books, 2007), xiv.

5   Lovelock, *The Ages of Gaia*, 8–10.

6   Lovelock, *The Ages of Gaia*, 41–61.

MODULE 4
THE AUTHOR'S CONTRIBUTION

KEY POINTS

- Although Lovelock's main aim in writing the book was to convey the core idea of a self-regulating,* living Earth to a general audience, he was aware that scientists might read about his Gaia hypothesis.*

- Lovelock's controversial Gaia hypothesis has changed the way life is viewed on Earth and eventually gave rise to the new academic field of geophysiology* or Earth system science*—a discipline founded on the principle that the Earth can be understood as a system of interactions in which life plays a significant role.

- It took more than a decade to develop the Gaia hypothesis beyond the mere idea of a self-regulating Earth; after discussions with scientists from several disciplines and with the help of a few close collaborators, the idea became the Gaia theory.*

Author's Aims

James E. Lovelock started writing *Gaia: A New Look at Life on Earth* in 1974. He believed that all living beings, including humans, are part of a community that is unconsciously keeping the Earth a comfortable place. He even started to feel that humans are like any other living organisms, with no particular rights but only obligations to the community of Gaia.[1] Lovelock wanted to share these ideas widely with the general public, not just with scientists who were not open to these rather controversial ideas at that time.

Lovelock conceived the idea of Gaia in 1965 while working at NASA.* He first published the idea in 1967 in the international

> **❝** The idea of the Earth as a kind of living organism, something able to regulate its climate and composition so as always to be comfortable for the organisms that inhabited it, arose in a most respectable scientific environment. It came to me suddenly one afternoon in 1965 when I was working at the Jet Propulsion Laboratory (JPL) in California. **❞**
>
> James E. Lovelock, *Gaia: A New Look at Life on Earth*

journal *Icarus*,* and then put the idea before his fellow scientists in 1971 in a presentation with the title "Gaia as Seen Through the Atmosphere."[2] Prestigious mainstream journals such as *Science* and *Nature*, however, were not ready to accept any paper on the Gaia hypothesis; "the very idea of detecting life on a planet by atmospheric analysis," Lovelock writes, "must have seemed outrageous to the conventional astronomers* and biologists* who reviewed our paper."[3]

In 1979, Lovelock finally published *Gaia*—a collection of all the ideas up to that point. He knew that his idea was significant and that he needed a concrete plan to translate the hypothesis into a testable scientific theory. Although he discussed the ideas with many scientists over the next decade, only a few supported his ideas and collaborated with him. Lovelock wanted to show his fellow scientists that Earth needed to be seen from top down, not the other way around; one can only see from space (he argued) that, compared to its dead neighbors, our blue planet Earth is living.

Approach

Lovelock writes his book, *Gaia: A New Look at Life on Earth* as a story of the discovery of life on the planet Earth. He begins the book by describing NASA's* space mission to search for life on other planets of the solar system. Lovelock addressed in an original way the core question of how to detect life on a distant world like Mars or Venus, suggesting that instead of testing soil, a planet's atmosphere can provide the necessary evidence if life indeed exists on its surface. This controversial idea departed from the scientific orthodoxy of that time, and Lovelock knew that without any evidence no one would believe him.

Between 1965 and 1975, he gathered as much supporting information as possible to develop his individual ideas into a working hypothesis. He thought the best way to describe his hypothesis was to write a book as a story of discovery. So in a series of interlinked chapters, Lovelock describes the atmosphere of the early Earth and how life evolved from tiny, single-celled organisms to more complex forms such as humans. Lovelock explains how living organisms interact with their nonliving environment to form a self-regulating entity, which he called Gaia. No author had previously offered such an in-depth analysis of the intricate yet complementary relationship between living and nonliving components within Gaia.

Contribution in Context

Although the core concept of a self-regulating, living Earth was not new, Lovelock's approach to researching it was original. At that time, no one was engaged in any research on the evolution of life and the evolution of the Earth as a single entity. When Lovelock started detailed research into the existing literature on life and its relationship with its environment, he could not find any. However, he stumbled upon some old notes and literature in which authors suggested that the Earth was alive. In the eighteenth century, for example, the geologist* James Hutton* said that the Earth was like an animal. A century later, the Austrian geologist Eduard Suess* introduced the word "biosphere",* which was further developed by the Russian geochemist* Vladimir I. Vernadsky,* who suggested that it can be regarded as the area on Earth in which energy is produced that sustains life.[4] While Lovelock was entirely ignorant of the related ideas of these earlier scientists, he later acknowledged those pioneers, referencing them in his second book *The Ages of Gaia* (1995).

NOTES

1   James E. Lovelock, *Gaia: A New Look at Life on Earth*, rev. ed. (Oxford: Oxford University Press, 2000), ix.

2   James E. Lovelock, *The Ages of Gaia: A Biography of Our Living Earth*, rev. ed. (Oxford: Oxford University Press, 1995), 8.

3   James E. Lovelock, *Homage to Gaia: The Life of an Independent Scientist*, rev. ed. (London: Souvenir Press Ltd., 2014), 250.

4   Lovelock, *The Ages of Gaia*, 8–10.

SECTION 2
IDEAS

MODULE 5
MAIN IDEAS

KEY POINTS

- A key part of the Gaia hypothesis* is a well-balanced and stable living condition known as homeostasis,* which results from interaction between living things and the environment.

- Homeostasis is maintained by a "cybernetic feedback mechanism"*—an automatic control system that makes adjustments in response to change.

- Lovelock warned that humans could break the feedback mechanism that sustains Gaia through too much change to the atmosphere and surface of the Earth.

Key Themes

James E. Lovelock's *Gaia: A New Look at Life on Earth* (1979) is a story about a search for life. Central to this book's story, the Gaia hypothesis suggests that the Earth constantly maintains the interaction of its physical, chemical, and biological environments so that organisms live and survive, evolving over time. This relative stability, called homeostasis, is fundamental for understanding the Gaia hypothesis.

In *Gaia*, Lovelock writes that the Earth maintains homeostasis through an automatic system called a cybernetic feedback mechanism.*[1] This system controls the temperature and the makeup of the Earth's atmosphere and oceans by making adjustments in response to change. The book says when the Earth's infant atmosphere was formed around 3.5 billion years ago, Gaia—the living Earth—shielded life on the planet by keeping the atmosphere suitable for living organisms.* The atmosphere also protected the planet from

> ** Earth's surface temperature is actively maintained
> comfortably for the complex entity which is Gaia, and
> has been so maintained for the most of her existence. **
>
> James E. Lovelock, *Gaia: A New Look at Life on Earth*

damaging cosmic radiation*—waves of energy from space—and
from bombardment by meteorites (rocks from space).[2]

The final theme of the book is sustainable living within Gaia.
Lovelock describes the relationship between humans and Gaia and
warns us of dire consequences if the Earth's internal feedback
mechanism fails. Such a failure could occur if humans cause excessive
modifications of the Earth's surface and atmosphere. The book
describes how humans are burning oil, gas, and coal more than ever
before, which adds more of the gas carbon dioxide* to the
atmosphere. This contributes to global warming* (an increase in
global temperatures) because of the greenhouse effect:* the higher
concentration of carbon dioxide and other greenhouse gases* in the
atmosphere traps more heat from the Sun.[3]

Exploring The Ideas

In *Gaia*, Lovelock explains how the composition of Earth's atmosphere
has changed over millions of years. Various atmospheric gases, primarily
nitrogen and oxygen, combined to create a homeostatic condition
(roughly, an equilibrium),* in which life is sustained despite a constantly
changing environment. Lovelock argues that Earth is different from its
neighboring planets.[4] The atmospheres of Mars and Venus are primarily
carbon dioxide gas and the life-supporting oxygen gas is almost absent.
In contrast, Earth's atmosphere has about 21 percent oxygen, which is
crucial for living animals and plants. Lovelock uses a conceptual model
to demonstrate that on an Earthlike planet that has no life and has
reached a state of chemical equilibrium, only a trace amount of life-

sustaining oxygen would be present.[5] But the oxygen content in the Earth's atmosphere has remained about 21 percent for the past 200 million years.[6] The consistent oxygen concentration suggests that an active control system is in place on the Earth's surface.

Lovelock calls this control system a cybernetic feedback mechanism.* To explain the concept, he gives examples of household appliances such as an electric oven, an iron, and a room heater. These appliances are equipped with a thermostat* that controls the desired temperature by switching on and off. A certain temperature is maintained through a feedback mechanism that tells the device to switch off when the temperature rises above the desired level. Similarly, the human body has feedback mechanisms to maintain its internal temperature, so that our bodily functions can keep us alive.[7] Lovelock emphasizes that a cybernetic feedback mechanism may exist at a planetary scale on Earth, where plants and animals have the capacity to regulate the Earth's climate.[8]

In the last two chapters, Lovelock turns to the theme of living sustainably within Gaia. He warns the reader that if humans keep modifying the surface and the atmosphere of the Earth then the delicate homeostatic system of the planet—which took several billion years to develop—will ultimately collapse and humans will face severe consequences. Lovelock expresses his great fear that the coexistence of life and Gaia is currently at stake because humans, for their own comfort, are continuously modifying the Earth's landscape and its delicate atmosphere. He writes in the book that "changes in the production rates of greenhouse gases may cause perturbations on the global scale"[9] that may seriously interfere with Gaia's state of homeostasis, and ultimately could threaten the existence of life on Earth.[10]

Language And Expression

In *Gaia*, Lovelock writes for a general audience, using simple language so that nonscientists can easily understand the strange and unfamiliar idea of a self-regulating, living Earth: "I tried to write this book so that a dictionary is the only aid needed."[11] According to the popular science journal *New Scientist*, "Lovelock writes beautifully. A book that is both original and well written is indeed a bonus."[12]

Lovelock has been successful; his simple but effective method ultimately brought him fame as a popular science writer. But on account of his simple writing style, lack of evidence, and the controversial nature of the topic, many scientists strongly criticized the book. The nonscientific nature of the book was emphasized by Lovelock's decision to name the self-regulating planet Gaia after the ancient Greek goddess of Earth[13] (a name suggested by the British novelist William Golding, author of *Lord of the Flies*, who was a resident of Lovelock's village).[14]

Despite the problems he had getting the idea recognized by scientists, Lovelock later said he never regretted the choice of name.

NOTES

1 See James E. Lovelock, "Cybernetics," in *Gaia: A New Look at Life on Earth*, by James E. Lovelock, rev. ed. (Oxford: Oxford University Press, 2000), 44–58.

2 Lovelock, *Gaia: A New Look at Life on Earth*, 59–77.

3 Lovelock, *Gaia: A New Look at Life on Earth*, 100–32.

4 James E. Lovelock, *Homage to Gaia: The Life of an Independent Scientist*, rev. ed. (London: Souvenir Press Ltd., 2014), 244.

5 Lovelock, *Gaia: A New Look at Life on Earth*, 30–46.

6 James E. Lovelock, *The Ages of Gaia*, rev. ed. (Oxford: Oxford University Press, 1995), 124.

7 Lovelock, *Gaia: A New Look at Life on Earth*, 49.

8 Lovelock, *Gaia: A New Look at Life on Earth*, 58.

9 Lovelock, *Gaia: A New Look at Life on Earth*, 113.

10 James E. Lovelock, *A Rough Ride to the Future* (London: Penguin Books, 2015), 75–111.

11 Brent F. Bauman, "The Feasibility of a Testable Gaia Hypothesis" (BSc thesis, James Madison University, 1998), 8.

12 Kenneth Mellanby, "Living with the Earth Mother," *New Scientist* 84 (1979): 41.

13 Toby Tyrrell, *On Gaia: A Critical Investigation of the Relationship between Life and Earth* (Princeton: Princeton University Press, 2013), 2.

14 Lovelock, *Homage to Gaia*, 255.

MODULE 6
SECONDARY IDEAS

KEY POINTS

- Trapping the Sun's heat, greenhouse gases* kept the early Earth's surface warm and comfortable, sustaining delicate life when the Sun's heat was much less intense than now.

- Self-regulating* mechanisms keep the level of oxygen in the atmosphere and the level of salt in the ocean relatively constant.

- The name Gaia and talk of a living Earth may initially have diverted the scientific community from giving the Gaia hypothesis* the attention it later received.

Other Ideas

In James E. Lovelock's *Gaia: A New Look at Life on Earth* there are several secondary ideas important to the development of his groundbreaking Gaia hypothesis:

- How are concentrations of different gases in Earth's atmosphere controlled?
- What is the importance of chemical reactions*—the effect of the combination of different chemicals—to the sustaining of life on Earth?
- What was the role of greenhouse gases in protecting the early Earth?
- Why is the sea not more salty, and what controls the salinity of seawater?

These secondary ideas, here stated as questions, help Lovelock explain and expand on the main themes. He explains, for example,

> **❝** The human species is of course a key development for Gaia, but we have appeared so late in her life that it hardly seemed appropriate to start our quest by discussing our own relationships within her. **❞**
>
> James E. Lovelock, *Gaia: A New Look at Life on Earth*

how a constant supply of energy from chemical reactions is needed if life is to be sustained; these reactions occur when two or more chemicals combine to create a different chemical. If a planet is in a state of chemical equilibrium* (balance), however, no reactions take place and no energy is produced.[1] Lovelock explains that "in such a world there is no source of energy whatever: no rain, no waves or tides, and no possibility of a chemical reaction which would yield energy."[2] Without the presence of any chemical reaction, no energy is produced on Mars or Venus—and these planets, our closest neighbors, are indeed lifeless.[3]

Exploring The Ideas

Life on Earth probably began as a simple, single-cell form under unsettled conditions of cosmic bombardment and active radioactivity*[4] (radioactivity occurs when atoms decay and release radiation*—roughly, waves of energy). Because of a lack of oxygen in the atmosphere, the surface of the Earth was exposed to the Sun's ultraviolet* radiation, which is not visible to the human eye. The atmosphere of the early Earth was filled with the gases carbon dioxide* and ammonia* (a combination of nitrogen and hydrogen). Lovelock suggests that these greenhouse gases kept the planet warm 3.5 billion years ago, when the Sun gave off 25 percent less heat than it does today.[5] Lovelock argues that 25 percent less heat from the Sun would imply an average surface temperature of well below zero degrees, which would cover the Earth's surface with snow and ice.

From the geological* records, however, it is known that for this period the climate of the Earth was never wholly unfavorable for life.[6] So while the Sun has steadily grown stronger over 3.5 billion years, Gaia has been sustaining life on Earth under a relatively stable climate.

Earth's atmosphere is made up primarily of nitrogen and oxygen, with small amounts of carbon dioxide and other minor gases. These gases, which play an important role in regulating the Earth's temperature and climate, are controlled by biota*—the animal and plant life in the environment. For instance, a constant level of oxygen in the atmosphere is possible because of an active biotic control. Green plants and algae* (a plantlike organism that lives in water) use light from the Sun to produce oxygen through a process known as photosynthesis,* creating nutrients and giving off oxygen in the process. This adds life-sustaining oxygen to the atmosphere—and yet, the level of oxygen remains constant at about 21 percent.

So how does Gaia control the Earth's atmosphere to keep an oxygen level suitable for life? The key to the control of oxygen is another gas, methane,* a combination of carbon and hydrogen, produced primarily by single-celled organisms— bacteria.* Although the atmosphere contains only trace amounts of methane, it is critical in controlling oxygen levels. Lovelock compares the function of methane with the function of glucose (a sugar) in a person's blood.[7] Glucose provides energy to cells in the human body, so maintaining fairly constant glucose levels is essential for healthy cells and, by extension, a healthy body. Similarly, methane controls the level of oxygen in the atmosphere by reacting with oxygen to form carbon dioxide and water.

Lovelock explains how the world's oceans are kept in balance. Rains and rivers dissolve salt from rocks and carry it over the land into the ocean, yet the salinity level in oceans has remained constant at around 3.4 percent for a very long time.[8] Lovelock argues that, throughout the history of the oceans' life, the salinity level could not have exceeded 6 percent as indicated by fossil records. At higher salinity levels, the life-forms in the ocean that we see today would have evolved very differently.[9] Lovelock argues that because salinity levels have been steady in the oceans, there must be a mechanism for the ocean to get rid of some of its salt.[10] He suggested that "excess salt accumulates in the form of evaporites* [deposits] in shallow bays, land-locked lagoons, and isolated arms of the sea, where the rate of evaporation* is rapid and the inflow [of salt] from the sea is one-way."[11]

Overlooked

Many scientists misinterpreted Lovelock's Gaia hypothesis. Lovelock writes in the revised edition that the book "is not for hard scientists. If they read it in spite of my warning, they will find it either too radical or not scientifically correct."[12] But scientists did read it as a scientific text, and reacted as Lovelock describes. The idea that living animals can change their environments through interactions with their nonliving counterparts was simply rejected by evolutionary biologists.* Other scientists called the Gaia hypothesis teleological* (in line with a belief that nature has a purpose) and goal-seeking*(based on repeated analyses looking for a particular answer). They argued that to self-regulate, organisms would need foresight and planning, which was impossible.[13]

Lovelock commented that "they see Gaia as metascience [beyond science], something like religious faith, and therefore from their deeply held materialistic beliefs, something to be rejected."[14] Two of Lovelock's supporters in the United States, the biologist* Stephen Schneider* and the environmental scientist Penelope J. Boston,* said in their book *Scientists on Gaia* (1993)[15] that the Gaia hypothesis wrongly attracted the most attention from theologians* who study religious ideas, usually through scripture. The turn of phrase "living Earth" and the name "Gaia" may have diverted the attention of the scientific community away from a serious analysis of the hypothesis and its implications.

NOTES

1 See James E. Lovelock, "The Recognition of Gaia," in *Gaia: A New Look at Life on Earth*, by James E. Lovelock, rev. ed. (Oxford: Oxford University Press, 2000), 32.

2 Lovelock, *Gaia: A New Look at Life on Earth*, 33.

3 Lovelock, *Gaia: A New Look at Life on Earth*, 30–43.

4 See James E. Lovelock, "In the Beginning," in *Gaia: A New Look at Life on Earth*, by James E. Lovelock, rev. ed. (Oxford: Oxford University Press, 2000), 15.

5 Lovelock, *Gaia: A New Look at Life on Earth*, 18.

6 Lovelock, *Gaia: A New Look at Life on Earth*, 18.

7 Lovelock, *Gaia: A New Look at Life on Earth*, 67.

8 Kate Ravilious, "Perfect Harmony," The *Guardian*, April 28, 2008, accessed December 30, 2013, http://www.theguardian.com/science/2008/apr/28/scienceofclimatechange.biodiversity.

9 Lovelock, *Gaia: A New Look at Life on Earth*, 86.

10 James E. Lovelock, *The Ages of Gaia: A Biography of our Living Earth*, rev. ed. (Oxford: Oxford University Press, 1995), 99–107.

11 Lovelock, *Gaia: A New Look at Life on Earth*, 91.

12 Lovelock, *Gaia: A New Look at Life on Earth*, xii.

13 James E. Lovelock, *Homage to Gaia: The Life of an Independent Scientist*, rev. ed. (London: Souvenir Press Ltd., 2014), 264.

14 Lovelock, *Gaia: A New Look at Life on Earth*, xii.

15 Stephen H. Schneider and Penelope J. Boston, eds., *Scientists on Gaia* (Cambridge: The MIT Press, 1993), 433.

MODULE 7
ACHIEVEMENT

KEY POINTS

- Lovelock wrote *Gaia: A New Look at Life on Earth* for a general audience because he believed, correctly, that the scientific community would not take it seriously.

- His warning about the danger of damaging the self-regulating* mechanisms of the Earth is directly related to today's concern about climate change.*

- The Gaia hypothesis* has led to the establishment of the interdisciplinary field of study called Earth system science.*

Assessing The Argument

James Lovelock's first book, *Gaia: A New Look at Life on Earth*, received a mixed reception when it was published in 1979. Lovelock later wrote in his autobiography that "its publication completely changed my life and the fall of mail through my letterbox increased from a gentle patter to a downpour, and has remained high ever since."[1] The main interest in Gaia came from the general public, philosophers, and religious leaders; only a third of the letters were from scientists.[2]

Although he did not write the book as a science text for specialists, Lovelock expected that some scientists would read it. Indeed they did—and many rejected the Gaia hypothesis. After the work's publication, Lovelock's scientific colleagues asked him why he reported the ideas about Gaia hypothesis in a book and not in peer-reviewed scientific journals.[3] He responded that when he published the first scientific article on Gaia in the early 1970s, the opposition,

> ❝ Now most scientists appear to accept Gaia theory
> and apply it to their research, but they still reject the
> name Gaia and prefer to talk of Earth System Science,
> or Geophysiology, instead. ❞
>
> James E. Lovelock, *Gaia: A New Look at Life on Earth*

mainly from evolutionary biologists,* was so strong that he believed
the editors of prestigious journals such as *Science* or *Nature* would
reject such articles.[4]

The situation changed in the 1990s, however, and Gaia papers
became easier to publish, even in prestigious journals. Lovelock
commented that the early rejection by biologists and the conservatism
of editors of some scientific journals helped him realize the potential
of the hypothesis in stimulating later scientific debates and discovery.[5]

Achievement In Context

Lovelock's Gaia hypothesis was born in the mid-1960s, when the
United States and the Soviet Union* were engaged in a serious battle
over space exploration, meaning that there was a significant focus on
space. His work at NASA* on scientific experiments for life detection
on Mars led him to think about why Earth was different from its
neighbors, Mars and Venus.

Since the work's publication in 1979, the Gaia hypothesis has
been actively debated in scientific circles. One result of that debate
was the development of a mathematical model called "Daisyworld,"*[6]
which ecologists use to test the role of biodiversity* (the richness of
species in a specific place) and stability of ecosystems* (habitat, the
organisms that live in it, and the interactions between the two) in
sustaining a healthy living environment.[7]

In his book, Lovelock also explains how the delicate self-
regulating system of Earth can collapse as a result of excessive

modification of the Earth's surface and atmosphere. In the 1970s, when the Gaia hypothesis predicted global changes, there was no evidence. Today, it is widely accepted that changes in atmospheric greenhouse gases* such as carbon dioxide are affecting the climate on a global scale.[8] These changes are called "anthropogenic,"* meaning they were caused by human activity. Lovelock warned about the danger of anthropogenic global warming* through the greenhouse effect* more than 30 years ago. The Gaia hypothesis is directly related to today's concerns about climate change.

Limitations

Gaia was published in 1979 in nonscientific and accessible language, and was received very well by the general public but rejected by some scientists, notably evolutionary biologists.[9] In addition to disagreeing on matters of science, scientists felt the storytelling approach and association of the book with a mythological character meant it could not be taken seriously.[10] Lovelock acknowledged that the "Gaia book was hypothetical, and lightly written—a rough pencil sketch that tried to catch a view of the Earth seen from a different perspective."[11] He also responded to criticisms from scientists by stating "I wrote this book when we were only just beginning to glimpse the true nature of our planet and I wrote it as a story of discovery."[12]

Another factor that limited the consideration of *Gaia* was that the very idea of a self-regulating Earth was so broad that it did not fit into a single, traditional academic discipline, such as geology,* biology,* or physics.*[13] At that time, taking an interdisciplinary approach to science was not yet common. Today, the Gaia hypothesis has moved from being "a metaphor,* not a mechanism,"[14] as the US evolutionary biologist Stephen J. Gould* understood it, to become the heart of a new interdisciplinary field of study: Earth system science. Lovelock's Gaia hypothesis was not limited to any particular time or place, and has inspired many academic disciplines, including ecology* (the branch of biology that looks at how organisms relate to one another and to their physical surroundings), marine biology* (the study of life in the oceans), and climate science.*[15]

NOTES

1 James E. Lovelock, *Homage to Gaia: The Life of an Independent Scientist*, rev. ed. (London: Souvenir Press Ltd., 2014), 264.

2 Lovelock, *Homage to Gaia*, 264.

3 James E. Lovelock, *The Ages of Gaia: A Biography of Our Living Earth*, rev. ed. (Oxford: Oxford University Press, 1995), xiii–xxii.

4 Lovelock, *The Ages of Gaia*, xv.

5 Lovelock, *The Ages of Gaia*, xiv–xv.

6 Andrew J. Watson and James E. Lovelock, "Biological Homeostasis of the Global Environment: The Parable of Daisyworld," *Tellus* 35B (1983): 286–89.

7 James E. Lovelock, *The Vanishing Face of Gaia: A Final Warning* (London: Penguin Books, 2010), 115.

8 James E. Lovelock, *Gaia: A New Look at Life on Earth*, rev. ed. (Oxford: Oxford University Press, 2000), 113.

9 Lovelock, *Homage to Gaia*, 264.

10 Lovelock, *Gaia: A New Look at Life on Earth*, xi.

11 Lovelock, *The Ages of Gaia*, 11.

12 Lovelock, *Gaia: A New Look at Life on Earth*, viii.

13 Lovelock, *Gaia: A New Look at Life on Earth*, xii–xiii

14 Stephen J. Gould, "Kropotkin Was No Crackpot," *Natural History* 106 (1997): 12–21.

15 Lovelock, *Homage to Gaia*, 241–79.

MODULE 8
PLACE IN THE AUTHOR'S WORK

KEY POINTS

- The Gaia hypothesis* has been the main focus of Lovelock's work for half a century. The idea of self-regulating Earth* was first presented at an academic conference in the mid-1960s.

- After *Gaia* was published, Lovelock and some collaborators began a decade-long effort to develop the concept into a broader Gaia theory* that would be testable and more scientifically thorough.

- Because Gaia did not fit entirely into any existing academic discipline, the new academic field of Earth system science* emerged.

Positioning

James E. Lovelock, the author of *Gaia: A New Look at Life on Earth*, has always cherished the freedom to follow his own ideas; for this reason, he never associated himself for long with a particular research or academic organization. This is also a reflection of his view that scientific inquiry is a form of direct engagement with the world.[1]

As an independent scientist and inventor, Lovelock invented a number of scientific devices, with the electron capture detector*—a device used to detect and measure atmospheric gases such as chlorofluorocarbons* ("CFCs")—being the most important. Lovelock writes that "electron capture detectors were undoubtedly the most valued of the trade goods which enabled me to pursue my quest for Gaia through the various scientific disciplines, and indeed to travel literally around the Earth itself."[2] Although this invention

> **❝** As a scientist, I submit wholly to scientific discipline and this is why I sanitized my second book, *The Ages of Gaia*, and hopefully made it acceptable to scientists. As a man, I also live in the gentler world of natural history, where ideas are expressed poetically and so that anyone interested can understand, and that is why this book [*Gaia*] remains almost unchanged. **❞**
>
> James E. Lovelock, *Gaia: A New Look at Life on Earth*

has nothing to do with the Gaia hypothesis, it gave him access to NASA's* space research laboratory, where the initial thoughts of this hypothesis* were seeded, later bringing him fame and recognition worldwide.

After publication of the first Gaia book in 1979, Lovelock's career took a slightly different turn. Other than concentrating on scientific invention, he became much more engaged in research on Gaia and in writing books and articles in scientific journals. The main focus of his research was further development of his Gaia hypothesis. Over the course of more than a decade, he consulted scientists from several disciplines as he sought to expand the Gaia hypothesis beyond the relatively simple idea of a self-regulating Earth. With the help of collaborators he developed the hypothesis into a theory* able to explain how human interventions in the environment such as the emission of greenhouse gases* like carbon dioxide* and methane* would interfere with the Earth's homeostatic* condition, endangering its delicate climate system.[3] "Homeostatic" here refers to an inclination toward balance, according to which living things hold themselves despite environmental change.

Throughout his career, Lovelock has written more than 200 scientific articles and published 10 books on Gaia. *Gaia: A New Look at Life on Earth*, his first book, remains his most famous.

Integration

Lovelock spent the first part of his professional career in medical research.[4] He conceived the main ideas of a self-regulating Earth at the age of 45 while working at NASA;* these developed into the Gaia hypothesis in the mid-1960s. Since then, the hypothesis has been Lovelock's primary area of research and publication. In his mid-90s at the time of writing, Lovelock is still active in Gaian research. His latest book on the subject, *A Rough Ride to the Future*, was published in 2015.[5]

Lovelock writes "the quest for Gaia has been a battle all the way."[6] But although the hypothesis has been fiercely debated in various scientific circles, even its most persistent critic, the US Earth scientist* James Kirchner,* noted that it has prompted many other scientists to form hypotheses of their own.[7] (Earth science is a discipline drawing on fields as diverse as geology,* chemistry,* biology,* and climatology, conducted to understand our planet's deep history and functioning as a system). Indeed, through collaboration with several scientists, in particular Lynn Margulis* and Andrew Watson,* the hypothesis evolved into a testable theory.

The value of any scientific theory is judged by the accuracy of its predictions.[8] Lovelock's Gaia theory made 10, among which were the lifelessness of Mars, and the critical roles of microorganisms and biodiversity* in the regulation of the planet's climate. So far, eight of these predictions have been confirmed or, at least, become generally accepted by scientists.[9] In his book *On Gaia* (2013), the environmental scientist and Gaian critic Toby Tyrrell* writes that "Lovelock was correct [that life has altered the Earth] both in terms of biological control over aspects of seawater chemistry, and also in terms of the composition of gases in the atmosphere."[10]

The Gaia theory, which draws on many academic disciplines— from astrobiology* (the branch of science concerned with life beyond Earth) to ecology*—has also made significant contributions to scientific research and discovery in many fields such as astrophysics* (the branch of astronomy concerned with the study of the physical characteristics of stars, galaxies, planets, and so on), biology, and Earth science. Finally, the concept and applications of Gaia theory have led to the development of an entirely new, interdisciplinary academic field, Earth system science.[11]

Significance

Lovelock's Gaia hypothesis has influenced many academics, scientists, and politicians over last 40 years. The former Czech president Vaclav Havel said, for example, that "according to the Gaia hypothesis, we are parts of a greater whole. Our destiny is not dependent merely on what we do for ourselves but also on what we do for Gaia as a whole."[12] But the hypothesis was not easily accepted by the scientific community.

Although Lovelock is well known for his scientific inventions in his field, it was the Gaia hypothesis that made him world famous. *Gaia: A New Look at Life on Earth* was bold and radical enough to spark a debate in the scientific community that has continued for decades— demonstrating the significance of the work in contemporary science.

The scientific database ScienceDirect* returns 146 books and 373 scientific articles in response to a search for the phrase "Gaia hypothesis." Scientific articles are still being published on the topic today. In 2006, the world's oldest geological professional body, the Geological Society of London, awarded Lovelock the Wollaston medal,* its highest honor, for his lifelong contribution to the study of the Earth.[13]

NOTES

1   John Gray, "James Lovelock: A Man for All Seasons," *New Statesman*, March 27, 2013, accessed December 21, 2013, http://www.newstatesman.com/culture/culture/2013/03/james-lovelock-man-all-seasons.

2   See James E. Lovelock, "Preface," in *Gaia: A New Look at Life on Earth*, by James E. Lovelock, rev. ed. (Oxford: Oxford University Press, 2000), xvii.

3   James E. Lovelock, *The Vanishing Face of Gaia: A Final Warning* (London: Penguin Books, 2010), 23–45.

4   James E. Lovelock, *Homage to Gaia: The Life of an Independent Scientist*, rev. ed. (London: Souvenir Press Ltd., 2014), 69–104.

5   James E. Lovelock, *A Rough Ride to the Future* (London: Penguin Books, 2015).

6   Lovelock, *Homage to Gaia*, 278.

7   James W. Kirchner, "The Gaia Hypothesis: Conjectures and Refutations," *Climatic Change* 58 (2003): 21.

8   Lovelock, *The Vanishing Face of Gaia*, 116–17.

9   Lovelock, *The Vanishing Face of Gaia*, 116.

10 Toby Tyrrell, *On Gaia: A Critical Investigation of the Relationship between Life and Earth* (Princeton: Princeton University Press, 2013), 202.

11 James E. Lovelock, "Geophysiology, the Science of Gaia," *Reviews of Geophysics* 27 (1989): 215–22.

12 Lovelock, *Gaia: A New Look at Life on Earth*, x.

13 Lovelock, *Rough Ride*, 77. This reference gives the wrong year (2003) for the award. The award was given in 2006. See "Wollaston Medal Citation," James Lovelock official website, accessed March 4, 2016, http://www.jameslovelock.org/page7.html.

SECTION 3
IMPACT

MODULE 9
THE FIRST RESPONSES

KEY POINTS

- The most important criticism was that Lovelock's Gaia hypothesis* was "scientifically untestable," simply teleological* (founded on a belief that nature has a purpose), and goal-seeking* (based on repeated analyses looking for a particular answer).

- Lovelock responded to criticisms of the Gaia hypothesis by writing his second book, *The Ages of Gaia*—written specifically for scientists.

- SWith evidence that organisms could indeed regulate their own environments, serving to change the status of the hypothesis to that of a testable theory, Gaian principles moved closer to acceptance by the scientific community.

Criticism

James E. Lovelock's Gaia hypothesis initially met with fierce disapproval. One of the strongest criticisms came from Heinrich Holland,* a geochemist* at Harvard University, who rejected the idea that living organisms controlled the regulation of Earth's atmosphere and its climate (geochemistry is the study of the chemical composition of, and chemical changes in, the solid matter of the Earth). Holland insisted climate was controlled by geochemical and geophysical* processes, and no life was actively involved.[1]

Damaging criticism also came from an American biologist, Ford Doolittle,* who reviewed the Gaia hypothesis in the journal *CoEvolution Quarterly* in 1981. Doolittle argued that there was no evidence that individual organisms could provide a cybernetic feedback

> 66 The critics took their science earnestly and to them
> mere association with myth and storytelling made it
> bad science ... I have tried ... [to convince scientists]
> both by rewriting my second book, *The Ages of Gaia*,
> so that is specifically for scientists, and leaving this
> book [*Gaia*] as it was. 99
>
> James E. Lovelock, *Gaia: A New Look at Life on Earth*

mechanism*—a physical, biological, or social response within a system that influences the continued activity or productivity of that system—as the hypothesis proposes.[2] He concluded that the Gaia hypothesis was an unscientific theory without any explained mechanism.

Another evolutionary biologist, Britain's Richard Dawkins,* criticized the Gaia hypothesis in *The Extended Phenotype: The Gene as the Unit of Selection* (1982) by arguing that organisms could not act as a unified group because this would require foresight and planning.[3] Echoing Doolittle's comments, he also rejected the idea that a cybernetic feedback mechanism could stabilize the system. Dawkins insisted that there was no way for evolution by natural selection to lead to altruism* (selfless behavior) on a global scale, as was proposed by Lovelock.[4] Dawkins argued that the Gaia hypothesis was teleological and goal-seeking.

The American evolutionary biologist Stephen J. Gould* described the Gaia hypothesis as a metaphorical* description of Earth's processes—that is, a figure of speech not to be taken literally;[5] in 1989, meanwhile, the Earth scientist* James Kirchner* wrote that "Gaia, in its different guises, is a mixture of fact, theory, metaphor, and wishful thinking."[6] For him, the Gaia hypothesis was not scientifically testable and was rather misleading in its suggestion that the Earth's environmental conditions have somehow been altered to meet the needs of the organisms.[7] The normal understanding of natural selection,* one of the fundamental principles of evolutionary theory, is that it is a process that allows organisms to become better adapted to their changing environments.

Despite Lovelock's declaration that the book was meant for a general audience, scientists criticized Lovelock for not using scientific language.[8] Others said the book was a mere expression of a religious faith or modern spiritual fantasy.[9]

Responses

Lovelock responded to criticisms from fellow scientists in various ways, saying critics did not understand, or that they were judging *Gaia* by the wrong standards. He responded very actively to one particular criticism, that the Gaia hypothesis was not testable. Lovelock's response was to launch a decade-long effort to develop his hypothesis into a scientifically rigorous theory that might become more widely accepted. The result was the Gaia theory.*

In 1983, Lovelock and his then postgraduate student, Andrew Watson,* developed an experimental computer model called "Daisyworld"*[10]—an imaginary Earthlike planet with a sun and a simple ecosystem* of two daisy species, black and white, which compete for space as they grow (an ecosystem can be understood, roughly, as a habitat and the organisms that survive in that habitat). The temperature of the planet is influenced by the proportion of ground covered by the two types of daisy. As the brightness of the sun increases over time, as has happened in reality with our own Sun, the planet becomes increasingly covered by white daisies. These reflect back more of the sun's warmth than the black daisies, and hence keep the planet cool.[11]

The model supported the idea that the temperature of an environment could be regulated by two competing species of plants living in the same environment so that ultimately an optimal temperature is reached. The Daisyworld model was accepted by many scientists, and particularly by mathematicians. One supporter was the mathematician Peter Saunders,* who commented that the model was worthy of study. Another supporter, the British Earth system scientist* Timothy Lenton,* has published a number of scientific papers on the implications of the Gaia theory and its mathematical basis.

In addition to developing the Gaia theory, Lovelock wrote a number of books on Gaia, including *The Ages of Gaia: A Biography of our Living Earth* (1988),[12] in which he explained his ideas using scientific language and responded to criticisms about his first book. Many scientific meetings and conferences took place to discuss the Gaia theory. For example, in 1994, scholars from many different disciplines discussed different aspects of Gaia theory and Earth's regulatory processes at a meeting titled "The Self-Regulating Earth."[13]

Conflict And Consensus

Lovelock's Gaia theory has become more complex and open to modification over the last four decades. Gaian thinking evolved from the provocative hypothesis that life controls planetary conditions for its own benefit, to a more robust and sophisticated theory that positions life as a key player in shaping the planet Earth.[14]

The concept of a cybernetic feedback mechanism for the earth's atmospheric regulation has been relevant to the debate around global warming,* as highlighted in 2001 by the Amsterdam Declaration on Global Change.* The environmental dimensions of Gaia theory revolve around two main concepts: the consequences of human-driven disruptions of the biosphere,* and the implications of habitat destruction and fragmentation of the Earth's ecosystems. While small-scale disturbances can be absorbed by the biosphere, large-scale disruptions sooner or later trigger far-reaching and uncontrollable consequences in the global climate.[15]

Gaian thought has also been addressed by sociologists (those studying society and social behavior). The sociologist Eileen Crist* and the ecologist* H. Bruce Rinker* have written: "The anthropogenic* amplification of the greenhouse effect* underway is rapid and large enough that it may unleash positive feedback … positive feedback, in turn, can trigger runaway heating. Such an eventuality will not only cause widespread human suffering, it will transform the Earth into a biological wasteland."[16]

In that sense, the Gaia hypothesis can be seen as a "game changer" because it has essentially shifted people's perception toward a more sustainable Earth.

While no consensus has been reached on whether the planet Earth is a self-regulating, living entity, Lovelock's Gaia and its central ideas, particularly the role of microorganisms in regulating Earth's atmosphere, are widely discussed in scientific work. Lovelock writes in his autobiography "if they [scientists] must reject Gaia as the name of their new science I hope that they will choose 'Earth System Science' as a sensible alternative."[17]

NOTES

1   James E. Lovelock, *The Vanishing Face of Gaia: A Final Warning* (London: Penguin Books, 2010), 111–12.

2   W. Ford Doolittle, "Is Nature Really Motherly?," *CoEvolution Quarterly* 29 (1981): 60.

3   See Richard Dawkins, *The Extended Phenotype: The Gene as the Unit of Selection* (Oxford: Oxford University Press, 1982), 1–307.

4   See Dawkins, *The Extended Phenotype*, 234–36.

5   Stephen J. Gould, "Kropotkin Was No Crackpot," *Natural History* 106 (1997): 12–21.

6   James W. Kirchner, "The Gaia Hypothesis: Fact, Theory, and Wishful Thinking," *Climatic Change* 52 (2002): 391.

7   James W. Kirchner, "Gaia Hypothesis: Can It Be Tested?," *Reviews of Geophysics* 27, no. 2 (1989): 223.

8   See James E. Lovelock, "Preface," in *Gaia: A New Look at Life on Earth*, by James E. Lovelock, rev. ed. (Oxford: Oxford University Press, 2000), x–xi.

9   Eileen Crist and H. Bruce Rinker, "One Grand Organic Whole," *Gaia in Turmoil: Climate Change, Biodepletion, and Earth Ethics in an Age of Crisis* (Cambridge: The MIT Press, 2010), 7.

10 Andrew J. Watson and James E. Lovelock, "Biological Homeostasis of the Global Environment: The Parable of Daisyworld," *Tellus* 35B (1983): 286–89.

11 Toby Tyrrell, *On Gaia: A Critical Investigation of the Relationship between Life and Earth* (Princeton: Princeton University Press, 2013), 25.

12 James E. Lovelock, *The Ages of Gaia: A Biography of Our Living Earth*, rev. ed. (Oxford: Oxford University Press, 1995).

13 James E. Lovelock, *Homage to Gaia: The Life of an Independent Scientist*, rev. ed. (London: Souvenir Press Ltd., 2014), 274–75.

14 Eileen Crist and H. Bruce Rinker, "One Grand Organic Whole," in *Gaia in Turmoil: Climate Change, Biodepletion, and Earth Ethics in an Age of Crisis*, ed. Eileen Crist and H. Bruce Rinker (Cambridge: The MIT Press, 2010), 8.

15 Crist and Rinker, "One Grand Organic Whole," 11–12.

16 Crist and Rinker, "One Grand Organic Whole," 14.

17 Lovelock, *Homage to Gaia*, 278–79.

MODULE 10
THE EVOLVING DEBATE

KEY POINTS

- The Gaia hypothesis has contributed significantly to the field of climate science,* particularly in the deep understanding of global climate and the role of microorganisms in regulating the Earth's atmosphere.

- Over the years, the concept of a self-regulating Earth* has slowly come to influence contemporary science by inspiring scientists, researchers, students, and politicians.

- The impact of the Gaia hypothesis on academic research and policies has been widespread, including the emergence of the new discipline of Earth system science,* and the relevance to the current debate about anthropogenic* climate change* through global warming.*

Uses And Problems

James E. Lovelock has continued to write about Gaia since the publication of his *Gaia: A New Look at Life on Earth* in 1979. In his *The Vanishing Face of Gaia* (2010) he describes how from the early 1980s to the mid-1990s it was impossible to publish any paper on the subject in any mainstream journal.[1] Since then, however, Lovelock's controversial hypothesis has influenced academics, scientists, politicians, and the general public for many years. In a critical book about the Gaia hypothesis in 2013, the British Earth system scientist Toby Tyrrell* writes: "In the thirty years or so since its inception, this Gaia hypothesis has vigorously inspired, infuriated, and intrigued a whole generation of environmental scientists."[2] From the mid-1960s to the late 1980s, Lovelock's hypothesis was severely criticized by geologists,*

> **"** Gaia theory aims to be consistent with evolutionary biology and views the evolution of organisms and their material environment as so closely coupled that they form a single, indivisible, process. **"**
>
> Timothy M. Lenton, "Gaia and Natural Selection," *Nature*

evolutionary biologists,* and planetary scientists.* Lovelock took critics' comments on board, and transformed his hypothesis into the more sophisticated, scientifically testable Gaia theory; in an updated edition of *Gaia*, he writes: "Gaia hypothesis was a vague speculation before the blood was drawn to leave the … more scientifically acceptable Gaia theory. For this I am grateful to the critics."[3]

Lovelock's first book, *Gaia*, did not convince many scientists—but a few academics welcomed the ideas it presented. One of these was René Dubos,* a French American microbiologist, who wrote in a review article that he read it with "immense pleasure,"[4] accepting that the Earth without life would produce an atmosphere filled with carbon dioxide gas that cannot support life.

In 1988, Lovelock published his second book, *The Ages of Gaia*—a more technical version of the first book. *The Ages of Gaia* laid out the main ideas more scientifically, explaining, for example, the critical role of ocean algae* (a plantlike organism) in climate regulation. In 1987, Lovelock and three collaborators, led by the American climate scientist Robert Charlson,* published a paper on what they called the CLAW hypothesis.*[5] The hypothesis, which draws its name from the collaborator's initials (Charlson, Lovelock, Andreae,* and Warren),* refined the Gaian concept of a cybernetic feedback mechanism* by attributing it specifically to microscopic ocean organisms known as phytoplankton.* A search for the phrase "CLAW hypothesis" on the database ScienceDirect finds 33 papers and 16 books by numerous authors who have mentioned the subject.

Schools Of Thought

In the 1980s, Lovelock and the British marine and atmospheric scientist Andrew Watson* developed the digital model they called Daisyworld* in order to translate the hypothesis into a scientifically defensible theory through testable proof.

Toby Tyrrell writes that "as Lovelock, now in his nineties, has become less active, others have taken up the torch."[6] For instance, Timothy Lenton,* an Earth system scientist at the University of Exeter in England, has published more than 20 articles on Gaia theory since 1997. Lenton is one of Watson's students; in 2011, the two published a book, *Revolutions that Made the Earth*, which expands on the ideas of the Gaia hypothesis.[7] He had previously made a significant contribution in a review article on the Gaia theory published in the journal *Nature* in 1998 in which he argued that the process of natural selection*—as famously described by Charles Darwin*—could be seen as an integral part of Gaia.[8]

Gaian models suggest that we have to think about organisms and their environments as one whole to understand fully which traits (characteristics that can be passed from generation to generation) come to persist and dominate. Scientists and researchers from around the world recognized in the Amsterdam Declaration on Global Change* of 2001 that "the Earth system behaves as a single, self-regulating system comprised of physical, chemical, biological and human components"—an idea central to Lovelock's work, even if the Declaration does not include the word "Gaia." [9]

In Current Scholarship

Today, scientists from many disciplines recognize Gaia theory as one of the most influential theories in contemporary science. For instance, Gaia theory adds to Darwin's vision that the success of species depends upon coherent coupling between the evolution of the organisms and the evolution of their material environment.[10] Despite his persistent criticism of the hypothesis, the Earth scientist James Kirchner* has nevertheless described it as "a fruitful hypothesis generator"—an inspiration for other research—for its having "prompted many intriguing conjectures about how biological processes might contribute to planetary-scale regulation of atmospheric chemistry and climate."[11]

Today, supporters of Lovelock's Gaia hypothesis come from disciplines as diverse as astrobiology* (inquiry into life elsewhere to the Earth), biology,* Earth system science, ecology,* environmental science,* and climate science. With his students, Andrew Watson (codeveloper of the Daisyworld model) has developed other mathematical models to represent regulation of atmospheric composition through geological time—which is to say through very long periods.[12]

NOTES

1   James E. Lovelock, *The Vanishing Face of Gaia: A Final Warning* (London: Penguin Books, 2010), 111.

2   Toby Tyrrell, *On Gaia: A Critical Investigation of the Relationship between Life and Earth* (Princeton: Princeton University Press, 2013), ix.

3   See James E. Lovelock, "Preface," in *Gaia: A New Look at Life on Earth*, by James E. Lovelock, rev. ed. (Oxford: Oxford University Press, 2000), xv.

4   René Dubos, "Gaia and Creative Evolution," *Nature* 282 (1979): 154–55.

5   Greg Ayers and Jill Cainey, "The CLAW Hypothesis: a Review of the Major Developments," *Environmental Chemistry* 4 (2007): 366–74.

6   Tyrrell, *On Gaia*, 3.

7   Timothy M. Lenton and Andrew Watson, *Revolutions that Made the Earth* (Oxford: Oxford University Press, 2013).

8   Timothy M. Lenton, "Gaia and Natural Selection," *Nature* 394 (1998): 447.

9   James E. Lovelock, "The Living Earth," *Nature* 426 (2003): 769–70.

10 James E. Lovelock, "Geophysiology, the Science of Gaia," *Reviews of Geophysics* 27 (1989): 222.

11 James W. Kirchner, "The Gaia Hypothesis: Conjectures and Refutations," *Climatic Change* 58 (2003): 21.

12 James E. Lovelock, *The Vanishing Face of Gaia: A Final Warning* (London: Penguin Books, 2010), 105–22.

MODULE 11
IMPACT AND INFLUENCE TODAY

KEY POINTS

- Lovelock's Gaia hypothesis* is still a hot topic for scientific debate and discovery nearly 40 years after the book's publication.

- Some critics of the Gaia theory* say that its claim that living things have a measure of control over the environment cannot be squared with the evolutionary theory derived from the work of Charles Darwin,* according to which living things adapt to the environment.

- Other critics challenge the predictive power of Gaia theory, and suggest that life may actually destroy the planet, rather than save it.

Position

James E. Lovelock's radical Gaia hypothesis as set out in his *Gaia: A New Look at Life on Earth* has continued to live among contemporary scientists and academics as a provocative scientific topic. Even after 40 years, the hypothesis is still being reexamined and reinterpreted by scientists of various fields.[1]

The Gaia theory is highly interdisciplinary, encompassing many disciplines ranging from astrobiology* to ecology.* Over a period of three decades, this theory has made significant contributions to many pieces of scientific research and to discoveries in astrophysics,* biology,* Earth science* and ecology. The concept behind Gaia theory has led to the development of a new multidisciplinary subject, geophysiology,* also known as Earth system science,* which treats the Earth as an interlinked system and tries to get a

> ❝ Gaia theory proposes that organisms inflicting damage on their surroundings will eventually reap harsh consequences when feedback comes back to haunt them. We are currently experiencing such feedback in the form of climate change, ozone depletion, endocrine disruption, and desertification. ❞
>
> Eileen Crist and H. Bruce Rinker, "One Grand Organic Whole," *Gaia in Turmoil*

better understanding of the physical, chemical, biological, and human interactions that determine the planet's past, current, and future states.

Another noteworthy contribution Gaia theory has made to climate science* and environmental science* is that it helps us better understand the current debate about global warming* and climate change.* The discipline of ecological philosophy, also known as deep ecology,* a way of looking at world problems that unites thinking, feeling, spirituality, and action, has been very much influenced by the Gaia hypothesis. Similarly, the idea of eco-spirituality,* which connects the science of ecology with spirituality, has been inspired by Lovelock's work.

According to the contemporary critic of the Gaia hypothesis Toby Tyrrell*—a professor of Earth system science at the University of Southampton—the Gaia hypothesis has achieved a degree of scientific respectability.[2] But while it is now accepted gladly by some, it also continues to stimulate intense debate.[3] Another persistent critic of the hypothesis, the Earth scientist* James Kirchner,* has written research articles[4,5] in scientific journals disputing the theory's predictive power. The importance of the Gaia hypothesis in explaining several vital mechanisms operating at the planetary scale on the Earth (such as the regulation of atmospheric gases) has guaranteed its continuing relevance in modern science.[6]

Interaction

Scientists from disciplines such as geology,* evolutionary biology,* and planetary science* have expressed serious reservations about the idea of a self-regulating, living planet. The main challenge is that the Gaia hypothesis suggests biotic control of the Earth's environment, meaning control by the biota*—the plant and animal life in the environment. This conflicts with evolutionary theory, according to which organisms adapt to their own environments.[7] Charles Darwin's theory of evolution through natural selection* holds that organisms with traits favoring survival will tend to live to reproduce and pass on those qualities to their offspring, while those that do not survive to reproduce will become extinct.[8]

Tyrrell and others have argued that natural selection operates according to the simple rule that whatever works best in time and space will be favored, regardless of future implications for the wider ecosystem or global impact.[9] Furthermore, global biota are not a closely related family group, and it is not possible that cooperation takes place at a scale as large as the whole Earth.[10]

The Continuing Debate

In his 2013 book *On Gaia: A Critical Investigation of the Relationship between Life and Earth,* Tyrrell considered why the Gaia hypothesis has an enduring appeal for some scientists and members of the general public. First, he thinks that Gaia is big-picture science that offers answers to deep questions such as that of why the Earth has remained continuously habitable for so long.[11]

Second, it suggests mechanisms for how our planet can cope in the future as it continues to be affected by global warming. Both James Kirchner and Toby Tyrrell argue that in order to protect planet Earth as a life-support system, our understanding of its natural processes must be based on a correct view. In this context, Kirchner provides some examples that challenge the predictive power of Gaia theory. For instance, Gaia theory predicts that biological processes should tightly regulate the makeup of the Earth's atmosphere, but rates of carbon uptake by microbes have increased by only about 2 percent in response to a 35 percent rise in atmospheric carbon dioxide* gas since preindustrial times.[12] Lovelock argues that Earth's system of self-regulation is now being overwhelmed by anthropogenic* greenhouse gas* pollution.

Another example of how the Gaia hypothesis has affected the contemporary intellectual world is the development of an anti-Gaian concept known as the Medea hypothesis* by Peter Ward,* an American paleontologist*—a researcher in fossils—at the University of Adelaide in Australia.[13] Like the Gaia hypothesis, the Medea hypothesis is named after a character from ancient Greek mythology: in this case, Medea—the wife of the mythological hero Jason. Ward argues that life is self-destructive, and gives several examples of mass extinction* in Earth's history. Ward's argument—which is essentially the opposite of the Gaia hypothesis—is that life will cause its own end by warming the biosphere about 1 billion years from now.[14]

NOTES

1 Toby Tyrrell, *On Gaia: A Critical Investigation of the Relationship between Life and Earth* (Princeton: Princeton University Press, 2013), 1–6.

2 Tyrrell, *On Gaia*, 1–6.

3 Tyrrell, *On Gaia*, 3.

4 James W. Kirchner, "Gaia Hypothesis: Can It Be Tested?," *Reviews of Geophysics* 27, no. 2 (1989): 223.

5 James W. Kirchner, "The Gaia Hypothesis: Conjectures and Refutations," *Climatic Change* 58 (2003): 21–45.

6 Crispin Tickell, "Scientists on Gaia," *Financial Times* 2002, accessed December 23, 2013, http://www.crispintickell.com/page19.html.

7 Timothy M. Lenton, "Gaia and Natural Selection," *Nature* 394 (1998): 439–47.

8   Charles Darwin, *On the Origin of Species by Means of Natural Selection, or the Preservation of Favoured Races in the Struggle for Life* (London: John Murray, 1859).

9   Tyrrell, *On Gaia*, 34.

10   Tyrrell, *On Gaia*, 40.

11   Tyrrell, *On Gaia*, 1–6.

12   Kirchner, "The Gaia Hypothesis: Conjectures and Refutations," 21.

13   Peter Ward, *The Medea Hypothesis: Is Life on Earth Ultimately Self-Destructive?* (Princeton: Princeton University Press, 2009), 208.

14   Moises Velasquez-Manoff, "The Medea Hypothesis: A Response to the Gaia Hypothesis," *The Christian Science Monitor*, February 12, 2010, accessed December 28, 2013, http://www.csmonitor.com/Environment/Bright-Green/2010/0212/The-Medea-Hypothesis-A-response-to-the-Gaia-hypothesis.

MODULE 12
WHERE NEXT?

KEY POINTS

- The Gaia hypothesis* has influenced scientific theories and discoveries over the last four decades and is relevant to academic research in the subject of global climate change.*

- Today, climate scientists modeling Earth's future climate are considering the suggestions made by the Gaia hypothesis—an important impact of Lovelock's groundbreaking work.

- The ideas offered by *Gaia: A New Look at Life on Earth* have led to the development of an entirely new discipline: Earth system science.*

Potential

Thanks to his publications such as *Gaia: A New Look at Life on Earth*, James E. Lovelock has been recognized by popular American magazine *Rolling Stone* as one of the twentieth century's most influential scientists.[1] According to the text, written for a nonscientific audience, living organisms* and their physical environment form a complete entity that controls the Earth's atmosphere and climate. The idea of a self-regulating Earth* was so radical at that time that many scientists criticized the Gaia hypothesis as simply bad science. In 1988, Lovelock published his second book, *The Ages of Gaia*, specifically for scientists.

As the Gaia theory* started to receive attention not only from the general public but from politicians, academics, and scientists, Lovelock wrote five more books about the Gaia hypothesis.[2] According to a piece in the British *Daily Telegraph* newspaper, the book "could prove

> ❝ We need the people of the world to sense the real and present danger so that they will spontaneously mobilize and unstintingly bring about an orderly and sustainable withdrawal to a world where we try to live in harmony with Gaia. ❞
>
> James E. Lovelock, *The Revenge of Gaia*

to be one of the twentieth century's most important pieces of polemic" ("polemic" here refers to a strongly worded work written with the intention to persuade). [3]

"Today [the Gaia hypothesis] is probably more widely credited than ever," writes Toby Tyrrell* in his critical 2013 book *On Gaia*.[4] The hypothesis has the potential to affect the debate over the Earth's climate and how it is regulated. In *Gaia* and subsequent books, Lovelock emphasizes anthropogenic* (human-caused) global warming.* Increased concentration of greenhouse gases* (particularly carbon dioxide*) has been driven by industrialization and increased burning of oil, gas, and coal. As the greenhouse effect* increases, the atmosphere traps more heat, and the loss of light-reflecting surfaces, mainly ice and snow, means that more light (and heat) is absorbed and less reflected, threatening to create a positive feedback effect and trigger runaway heating.[5]

Under these circumstances, the Gaia hypothesis can offer scientists and policymakers a unique perspective from which current environmental problems can be viewed. Indeed, the Gaia hypothesis has already made significant contributions in understanding the issue of global warming.[6]

Future Directions

Lovelock writes in his *The Vanishing Face of Gaia: A Final Warning* (2010) that changes in the Earth's climate could lead to the disappearance of sensitive ecosystems,* and potentially endanger the existence of human life.[7] Lovelock strongly criticizes some climate scientists and politicians for failing to see the Earth as a self-regulating, living entity.[8] For a long time, they considered the Earth as a solid rock and did not consider biotic* (biological) interactions with the Earth's physical environments.[9] Recently, the Intergovernmental Panel on Climate Change (IPCC)* has considered new climate models, known as Earth system models,* that recognize the interaction of multiple factors in affecting the climate.

According to an article in the British *Guardian* newspaper, Lovelock "is not a doom-monger but a practical problem-solving man, with suggestions for alleviating the climate crisis at many levels."[10] Lovelock strongly believes that the concept of Gaia will develop further in the future. The Earth system scientist Timothy Lenton,* for instance, has been leading research on the Gaia hypothesis. In his recent book *The Vanishing Face of Gaia*, Lovelock claims that new technologies such as geoengineering*—deliberate large-scale intervention in the Earth's natural systems—will emerge to counteract climate change.

Summary

Many applications of the Gaia hypothesis have materialized since the publication of *Gaia: A New Look at Life on Earth* in 1979. Lovelock provided a number of interesting ideas in his book for further academic research. His Gaia theory has prompted some intriguing interpretations of the way biological processes could contribute to planetary-scale regulation of atmospheric chemistry and climate.[11] In *Gaia*, Lovelock argues that Earth is a self-regulating system. Several planetary mechanisms such as the regulation of atmospheric gases, atmospheric temperature, and the salt levels of oceans are explained in the book. Lovelock also explains how these systems are controlled by living organisms on Earth through a cybernetic feedback mechanism*—an automatic control system that makes adjustments in response to change. All these ideas described in Lovelock's *Gaia* have significantly contributed to academic research and applications in the field of ecology* and climate science* over the last 40 years.

The Gaia hypothesis is interdisciplinary in nature, drawing on the aims and knowledge of many different academic disciplines; it does not fall within a single one. Lovelock proposed the study of Earth systems within a new discipline, geophysiology,* or Earth system science.[12] Gaia's ideas were also considered by academics of other disciplines such as ecology, marine biology,* and climate science. One of the most remarkable applications of the Gaia hypothesis was the development of a mathematical model called Daisyworld*[13]— now used by ecologists to test the role of biodiversity* and stability of ecosystems for a healthy living environment.

A more recent application of the Gaia hypothesis is seen in school education.[14] A 2009 study in Brazil has found that the Gaia hypothesis can contribute to the understanding of human activities and contemporary environmental issues such as global warming and climate change in science education at school. The study has also found that the interdisciplinary nature of Lovelock's controversial Gaia hypothesis makes an interesting and effective tool for cross-disciplinary learning at school.

NOTES

1   Jeff Goodell, "James Lovelock, the Prophet," *Rolling Stone*, November 1, 2007, accessed December 23, 2013, http://www.rollingstone.com/politics/news/james-lovelock-the-prophet-20071101.

2   James E. Lovelock, *Homage to Gaia: The Life of an Independent Scientist* (2000), *Healing Gaia: The Practical Science of Planetary Medicine* (2001), *Gaia: Medicine for an Ailing Planet* (2005), *The Revenge of Gaia: Why the Earth is Fighting Back—and How we Can Still Save Humanity* (2007), and *The Vanishing Face of Gaia: A Final Warning* (2010).

3   James Flint, "Earth—The Final Conflict," The *Daily Telegraph*, February 6, 2006, accessed December 23, 2013, http://www.telegraph.co.uk/culture/books/3649909/Earth-the-final-conflict.html.

4   Toby Tyrrell, *On Gaia: A Critical Investigation of the Relationship Between Life and Earth* (Princeton: Princeton University Press, 2013), ix.

5   Eileen Crist and H. Bruce Rinker, "One Grand Organic Whole," in *Gaia in Turmoil*, ed. Eileen Crist and H. Bruce Rinker (Cambridge: The MIT Press, 2010), 1–20.

6   James E. Lovelock, *A Rough Ride to the Future* (London: Penguin Books, 2015), 85–103.

7   James E. Lovelock, *The Vanishing Face of Gaia: A Final Warning* (London: Penguin Books, 2010), 1–45.

8   James E. Lovelock, *The Revenge of Gaia: Why the Earth is Fighting Back—and How We Can Still Save Humanity* (London: Penguin Books, 2007), 61–83.

9   Lovelock, *Vanishing Face of Gaia*, 1–45.

10 Peter Forbes, "Jim'll Fix It," The *Guardian*, February 21, 2009, accessed December 23, 2013, http://www.theguardian.com/culture/2009/feb/21/james-lovelock-gaia-book-review.

11 James W. Kirchner, "The Gaia Hypothesis: Conjectures and Refutations," *Climatic Change* 58 (2003): 21.

12 James E. Lovelock, "Geophysiology, the Science of Gaia," *Reviews of Geophysics* 27 (1989): 215–22.

13 Andrew J. Watson and James E. Lovelock, "Biological Homeostasis of the Global Environment: The Parable of Daisyworld," *Tellus* 35B (1983): 286–89.

14 Ricardo Santos do Carmo, Nei Freitas Nunes-Neto, and Charbel Nino El-Hani, "Gaia Theory in Brazilian High School Biology Textbooks," *Science & Education* 18 (2009): 469–501.

GLOSSARY

GLOSSARY OF TERMS

Algae: plantlike organisms that can make food in the presence of sunlight by photosynthesis.

Altruism: a human behavior that displays a desire to help others selflessly.

American Geophysical Union: a nonprofit organization founded for the advancement and wider understanding of research in geophysics.

Ammonia: a strong, colorless gas composed of nitrogen and hydrogen with a characteristic pungent odor.

Amsterdam Declaration on Global Change: a declaration made by the scientific communities of four international global change research programs, recognizing that ever-increasing human modifications of the global environment have great implications for human well-being, in addition to the threat of climate change.

Anthropogenic: something caused or influenced by humans.

Apollo mission: a spaceflight program carried out by the National Aeronautics and Space Administration (NASA). Apollo astronauts became the first human beings on the Moon between 1969 and 1972.

Artificial satellite: a human-built object orbiting the Earth and other planets in the solar system.

Astrobiology: the science concerning the origin and evolution of life beyond Earth, in the universe.

Astronomer: a scientist who studies stars, planets, moons, comets, galaxies, and so on.

Astrophysics: the branch of astronomy that studies the physical characteristics and composition of celestial objects such as stars, moons, planets, and so on.

Atmospheric gases: gases present in the Earth's atmosphere such as oxygen, carbon dioxide, nitrogen, and so on.

Bacteria: single-celled, microscopic organisms that are found everywhere both inside and outside of human bodies. Some bacteria are helpful for human health but the majority are harmful.

Biochemist: a scientist with a qualification in biochemistry—the branch of science that studies chemical processes in living organisms.

Biodiversity: the diversity (variety) of species in any particular location.

Biology: the scientific study of living things such as plants and animals.

Biologist: a scientist who focuses on living organisms, including plants and animals.

Biosphere: the regions of the surface and atmosphere of the Earth or another planetary body that are occupied by living organisms.

Biota: living organisms such as animals and plants of a region, habitat, or geological period.

Carbon dioxide: a colorless, odorless, and incombustible gas that is commonly found in the Earth's atmosphere. A greenhouse gas, it is formed during respiration and the decomposition and combustion of organic matter.

CFCs: See chlorofluorocarbons.

Chemistry: the branch of science that studies chemical properties of substances and their reactions.

Chlorofluorocarbons: also known as CFCs, these are nontoxic and nonflammable chemical compounds containing atoms of carbon, chlorine, and fluorine. They are commonly used as coolants.

CLAW hypothesis: developed by Lovelock and three other scientists, the CLAW hypothesis proposes that microscopic organisms inhabiting the ocean surface are able to regulate their own population; the name is derived from the initials of the four scientists Robert Charlson, James Lovelock, Meinrat Andreae, and Stephen G. Warren.

Climate change: a long-term change in the planet's weather patterns such as a shift in average temperatures.

Climate science: the study of climate—the average condition of day-to-day weather measured over a considerably long period.

Continental drift: a theory proposed by German scientist Alfred Wegener that the continents drift and move position on Earth's surface over millions of years.

Cosmic radiation: an emission of cosmic rays made up of energetic, subatomic-sized particles arriving from outside the Earth's atmosphere.

Cybernetic feedback mechanism: a response within a mechanical, physical, biological, or social system that influences the continued activity or productivity of that system.

Daisyworld: in a digital model designed by Lovelock and his student Andrew Watson to provide testable evidence for the Gaia hypothesis, Daisyworld was a cloudless hypothetical Earthlike planet with a negligible atmospheric greenhouse. There were two daisy species on the planet—one black, one white. The ground covered by the black (dark) daisy reflected less light than clear ground; the ground covered by the white (light) daisy reflected more light than the bare ground.

Deep ecology: an all-inclusive approach that combines feeling, spirituality, thinking, and action in seeking to solve the world's problems. It involves moving beyond the self-centred nature of modern culture toward seeing human beings as part of the Earth.

Disequilibrium: a loss or lack of equilibrium or stability in which opposing forces or influences are not in balance with each other.

Earth science: a discipline drawing on fields as diverse as geology, chemistry, biology, and climatology conducted to understand our planet's deep history and functioning as a system.

Earth system models: models that look at the interactions of atmosphere, ocean, land, ice, and biosphere in order to gauge the state of regional and global climatic conditions under a wide variety of settings.

Earth system science: an interdisciplinary field of study that identifies the Earth as a unified system and looks for a way to understand the interaction of the environment (physical, chemical, and biological) and human activity to establish the state of the planet in the past, present, and future.

Earth system scientist: a scientist in the field of Earth system science.

Ecology: the branch of biology that studies the relationship of organisms to each other and to their physical environments.

Eco-spirituality: a philosophy based in a fundamental belief in the sacredness of nature, Earth, and the universe.

Ecosystem: a biological system made up of organisms found within a specific physical environment that interact both with the environment and also with each other.

Electron capture detector: a device used to detect trace amounts of chemical compounds in a sample.

Environmental science: a multidisciplinary science that looks at environmental conditions and their effect on organisms living in that environment.

Equilibrium: a state in which opposing forces or influences are balanced with each other.

Evaporation: the process by which a substance in a liquid state changes to a gas due to an increase in temperature or pressure or a combination of both.

Evaporites: a deposit of minerals, formed by the evaporation of salt water.

Evolution: a gradual change in the characteristics of a group of animals or plants that takes place over successive generations and explains the development of existing species from their dissimilar ancestors.

Evolutionary biology: a branch of biology that studies the evolution of organisms, especially molecular and microbial evolution. Other areas involve behavior, genetics, ecology, life histories, and development.

Gaia hypothesis: an idea that the physical and chemical condition of the surface of the Earth, of the atmosphere, and of the oceans has been and is actively made fit and comfortable by the presence of life itself—according to which, as James Lovelock puts it, "the entire surface of the Earth including life is a self-regulating entity."

Gaia theory: a theory derived from the Gaia hypothesis and substantially proved by a mathematical model developed by James Lovelock and Andrew Watson ("Daisyworld").

Gaian thinking: the philosophical position that all organisms on Earth, including humans, interact with each other to shape their living environment and keep it fit and comfortable—the way of thinking holistically inspired by the Gaia hypothesis.

Geochemist: a scholar of the chemical composition of and chemical changes in the solid matter of the Earth or other planetary bodies.

Geoengineering: an intentional, large-scale technological intervention in the Earth's natural systems. Geoengineering is often discussed as a techno-fix for combating global climate change.

Geology: the study of the origin, history, and structure of the Earth—the science dealing with the solid Earth and its rocks.

Geophysicist: a scholar of the various gravitational, magnetic, electrical, and seismic phenomena (such as earthquakes) that define a planet.

Geophysiology: the study of interactions among all living organisms on Earth. Geophysiology is a transdisciplinary environment for studying planetary-scale problems.

Geosciences: the sciences concerned with the Earth such as geology, geophysics, and geochemistry.

Global warming: a gradual increase in long-term average temperature of the Earth's atmosphere.

Goal-seeking: the process of calculating an output by performing various "what-if" analyses on a given set of inputs.

Greek mythology: myths and tales about the ancient Greeks. Greek mythology is concerned with mythological gods and heroes, various other mythological creatures, and the origins and importance of the ancient Greeks' cult and ritual practices.

Greenhouse effect: the natural process by which the Earth's atmosphere traps some of the Sun's energy, keeping the atmosphere warm enough to support life. It is referred to as the greenhouse effect because a greenhouse works in much the same way, trapping heat within itself.

Greenhouse gases: gases that contribute to the greenhouse effect by absorbing infrared radiation invisible to human eyes.

Homeostasis or homeostatic condition: a state of constancy in which living things hold themselves when the environments in which they reside are changing.

Humanism: a philosophical position that emphasizes human concerns; in its rational form, it serves as an alternative ethical system to those derived from religion.

Hypothesis: an idea or concept that is not proven scientifically but leads to further study or discussion. In science, a hypothesis needs to go through much testing before it can earn the status of a theory. In the nonscientific world, the words hypothesis and theory are often used interchangeably.

Icarus: a scientific journal dedicated to the field of planetary science.

Intergovernmental Panel on Climate Change (IPCC): a scientific intergovernmental body represented by all nations under the auspices of the United Nations.

Marine biology: the study of marine species or organisms that live in the ocean and other salt-water environments.

Mass extinction: a relatively sudden decrease in the diversity of animals or plants on a global scale. Throughout the history of life on Earth there have been mass extinctions and they generally occur in a short amount of geological time.

Medea hypothesis: the idea that life is its own enemy and that nearly all mass extinctions on Earth were caused by life itself. This hypothesis, put forward by American paleontologist Peter Ward, stands in complete contrast to Lovelock's Gaia hypothesis, which suggests that life itself sustains a comfortable living condition on

Earth. Gaia views the Earth as a "good mother," whereas Medea views the Earth as an "evil mother."

Metaphor: a word or phrase generally used to compare two objects, ideas, thoughts, or feelings that are different.

Methane: a colorless, odorless, and flammable gas. The major constituent of natural gas, methane is the simplest of the hydrocarbons. It is released during the decomposition of plant or other organic compounds. Methane is also considered a greenhouse gas.

NASA: the National Aeronautics and Space Administration. NASA is the United States government agency responsible for the civilian space program.

National Institute for Medical Research: a medical research institute situated near London, England. It was set up in 1913 by the UK's Medical Research Council.

Natural selection: the process through which evolutionary changes occur when individual organisms such as plants and animals that have certain characteristics achieve a greater survival or reproductive rate than other individuals in the same population. The more successful individuals then pass these useful inheritable characteristics on to their offspring.

Organism: an animal, plant, fungus, or bacterium that is a living biological entity.

Paleontology: the scientific study of plant and animal fossils.

Photosynthesis: the process by which green plants and other organisms use light energy (usually from sunlight) to create nutrients from carbon dioxide and water. Plants produce food by this process; oxygen is also released as a waste product.

Physics: the branch of fundamental science that studies the nature and properties of matter and energy.

Phytoplankton: tiny marine plants that, in a balanced ecosystem, provide food for a huge range of sea creatures including jellyfish, whales, and snails.

Planetary science: the study of planets including the Earth, moons, and planetary systems, particularly those belonging to the solar system.

Planetary scientist: a scholar of planets.

Radiation: energy transmitted in waves or in a stream of particles— heat and light from a fire, for example, is a form of radiation.

Radioactivity: emissions—radiation, or energy waves—produced by the decay of atoms.

Reductionism: the practice of seeing something complex as merely the sum of its parts, without considering how those parts may interact with each other.

ScienceDirect: a scientific database that offers peer-reviewed journal articles and book chapters in the sphere of science and medicine.

Self-regulating Earth: the position that all organisms interact with the Earth's air, water, and rocks in order to keep the planet fit and comfortable for life in a stable fashion.

Soviet Union: also known as the Union of Soviet Socialist Republics (USSR), a socialist state on the European continent. The USSR, in existence 1922–91, was a one-party state governed by the Communist Party. Moscow was its capital city.

Space exploration: the investigation of biophysical conditions beyond the boundary of Earth—space, stars, planets, comets, etc—using manned spacecraft, satellites or probes.

Sputnik: a Russian satellite launched into space in 1957. It was the first human-made object ever to leave the Earth's atmosphere.

Superorganism: an entity made up of many distinct organisms.

Teleological: relating to teleology—an approach that explains the reason or explanation for something in terms of its end, purpose, or goal.

Theology: the study of God and religious faiths.

Theory: a natural explanation for a group of facts or phenomena. Theories are coherent, predictive, systematic, and widely applicable. Theories generally undergo testing and improvement or modification as more information is generated so that the predictive capacity of a theory becomes greater over time.

Thermostat: an automatic apparatus for regulating temperature in an electrical device such as a kitchen oven.

Ultraviolet: part of the spectrum of light invisible to the human eye.

United Nations: an international institution founded in 1945 with the aim of promoting international cooperation, peace, and security.

Viking Mission to Mars: NASA's mission to Mars; the probes launched in 1975 were the first to land safely on the surface and return images.

Wollaston medal: the highest award in the field of geoscience given by the Geological Society of London. Geologists who have contributed significantly through excellent research in fundamental and/or applied aspects of geoscience may be honored with this award.

Working class: a socioeconomic term used to describe people in a social class marked by jobs that provide relatively low pay, require limited skill sets, and low educational requirements.

PEOPLE MENTIONED IN THE TEXT

Meinrat Andreae (b. 1949) is a German mineralogist, currently serving as the director of the biogeochemistry department at the Max Planck Institute for Chemistry in Germany.

Penelope J. Boston is a professor in the department of Earth and environmental science at New Mexico Tech University. Boston is well known for proposing that small jumping robots be sent to Mars to facilitate exploration.

Robert Charlson is a professor emeritus of atmospheric sciences and chemistry at the University of Washington. Charlson collaborated with James Lovelock and two other scientists, Meinrat Andreae and Stephen G. Warren, on the CLAW hypothesis.

Eileen Crist is a sociologist, and an associate professor of science and technology in society at Virginia Polytechnic Institute and State University (Virginia Tech).

Charles Darwin (1809–82) was an English naturalist and geologist. Darwin is best known for his theory of evolution by means of natural selection.

Richard Dawkins (b. 1941) is a British evolutionary biologist, a popular writer, and an outspoken atheist. Dawkins served as a professor at the University of Oxford and is now an emeritus fellow of New College, Oxford.

Ford Doolittle (b. 1941) is a biochemist who was born in the US state of Illinois. He is a professor at Dalhousie University in Canada and has been a long-term critic of Lovelock's Gaia hypothesis.

René Dubos (1901–82) was a French-born American microbiologist and a professor emeritus at Rockefeller University. Dubos was one of the earlier supporters of the main ideas of the Gaia hypothesis.

Nellie A. Elizabeth was James Lovelock's mother who worked as a personal secretary. She used to take Lovelock to the local library to borrow science fiction books to read.

Sidney Epton worked for Shell's Thornton Research Centre in the United Kingdom and collaborated with James Lovelock in the 1970s in developing the Gaia hypothesis.

William Golding (1911–93) was an English novelist who the Nobel Prize for Literature in 1983. A resident of Lovelock's village, he suggested the name "Gaia" for Lovelock's hypothesis.

Stephen J. Gould (1941–2002) was an American paleontologist and evolutionary biologist who spent most of his career teaching at Harvard University. Gould was very critical of Lovelock's Gaia hypothesis.

Václav Havel (1936–2011) was a Czech playwright, president of Czechoslovakia between 1989 and 1992 and of the Czech Republic between 1993 and 2003.

Dian Hitchcock is a philosopher who worked with James Lovelock at NASA, where they examined the atmospheric data from Mars and concluded that Mars was lifeless. A decade later, it was confirmed by missions to Mars that their conclusions were correct.

Heinrich Holland (1927–2012) was born in Germany, but settled in the United States. Holland was an emeritus professor at Harvard University and he made major contributions to the understanding of the Earth's geochemistry.

George Hutchinson (1903–91) was an American zoologist, best known for his ecological studies of freshwater lakes. Hutchinson was born in England and educated at Cambridge University. In 1928, he joined the faculty of Yale University to teach zoology and he spent most of his professional life there.

James Hutton (1726–97) was a highly influential geologist from Scotland whose ideas anticipated those of Lovelock.

Thomas H. Huxley (1825–95) was an English biologist, best known for his work on Charles Darwin's theory of evolution by natural selection. Huxley did more than anyone else to get Darwin's theory accepted by scientists and the general public.

John F. Kennedy (1917–63) also known as JFK, was the 35th president of the United States (1961–63) and the youngest man elected to the office. On November 22, 1963, he was assassinated in Dallas, Texas, becoming also the youngest president to die.

James W. Kirchner is currently a professor of Earth and planetary science at the University of California, Berkeley. Kirchner has been very interested in, if critical of, the Gaia hypothesis.

Yevgraf M. Korolenko was a nineteenth century Ukrainian philosopher and scientist. Korolenko was a learned man, although self-educated; he was familiar with the works of the great natural scientists of his time.

Timothy Lenton is professor of climate change and Earth system science at the University of Exeter. Throughout his career Lenton has been very interested in James Lovelock's controversial Gaia hypothesis and is considered a possible successor to Lovelock.

Thomas A. Lovelock was James Lovelock's father. In professional life, Tom Lovelock was an art dealer and had very strong feelings about nature.

Lynn Margulis (1938–2011) was a distinguished professor of geosciences at the University of Massachusetts, Amherst. Margulis collaborated with James Lovelock and together they developed the controversial Gaia hypothesis.

Mary Midgley (b. 1919) is an English moral philosopher whose special interest covers science, ethics, human nature, and animal rights. Midgley wrote in favor of a moral interpretation of Lovelock's Gaia hypothesis.

Vance Oyama (1922–98) was a biochemist who worked for NASA on the search for life on Mars. Oyama will be remembered for his pioneering life detection experiments on Apollo lunar samples.

H. Bruce Rinker is an ecologist and executive director of the Valley Conservation Council in Virginia.

Peter Saunders is emeritus professor of mathematics at King's College, London.

Stephen Schneider (1945–2010) was an American professor of environmental biology and global change at Stanford University. He was internationally recognized for his research, policy analysis, and outreach in the subject of global climate change.

Eduard Suess (1831–1914) was an Austrian geologist who contributed greatly to the knowledge of his field. Suess is credited with generating many of the concepts that led to the theory of plate tectonics (the movement of the Earth's crust) and paleogeography (the study of ancient land masses).

Toby Tyrrell is a professor of Earth system science at the University of Southampton. Tyrrell is well known for his critical review of the Gaia hypothesis.

Vladimir I. Vernadsky (1863–1945) was a renowned Russian mineralogist and geochemist. Vernadsky is considered the founder and father of modern geochemistry.

Jules Verne (1828–1905) was a nineteenth-century French novelist and poet. He is the author of *Around the World in Eighty Days* and *20,000 Leagues Under the Sea*.

Peter Ward (b. 1949) is an American paleontologist and currently a professor at the University of Adelaide in Australia. He is known for his anti-Gaian hypothesis, *The Medea Hypothesis*, published in 2009.

Andrew Watson (b. 1952) is a British marine and atmospheric scientist, currently a professor at the University of Exeter. Watson was a PhD student of James Lovelock; together, they developed the computer model "Daisyworld" to provide proof that organisms can regulate their environment.

Alfred Wegener (1880–1930) was a polar researcher, geophysicist, and meteorologist born in Berlin, Germany. Wegener became famous for his groundbreaking theory of continental drift that, for the first time, suggested that all continents were slowly moving around the Earth.

H. G. Wells (1866–1946) was a British novelist. The author of *The Time Machine* and *The War of the Worlds*, Wells is best known for his science fiction novels.

WORKS CITED

WORKS CITED

Ayers, Greg and Jill Cainey. "The CLAW Hypothesis: A Review of the Major Developments." *Environmental Chemistry* 4 (2007): 366–74.

Bauman, Brent F. "The Feasibility of a Testable Gaia Hypothesis." BSc Thesis, James Madison University, 1998.

Carmo, Ricardo S. do, Nei Freitas Nunes-Neto, and Charbel Nino El-Hani. "Gaia Theory in Brazilian High School Biology Textbooks." *Science & Education* 18 (2009): 469–501.

Crist, Eileen, and H. Bruce Rinker. "One Grand Organic Whole." In *Gaia in Turmoil: Climate Change, Biodepletion, and Earth Ethics in an Age of Crisis*, edited by Eileen Crist and H. Bruce Rinker, 3–20. Cambridge: The MIT Press, 2010.

Darwin, Charles. *On the Origin of Species by Means of Natural Selection, or the Preservation of Favoured Races in the Struggle for Life*. London: John Murray, 1859.

Davis, David. "A Few Thoughts on the Apocalypse." *The Spectator*, February 25, 2009. Accessed December 21, 2013. http://www.spectator.co.uk/features/3387731/a-few-thoughts-on-the-apocalypse/.

Dawkins, Richard. *The Extended Phenotype: The Gene as the Unit of Selection*. Oxford: Oxford University Press, 1982.

Doolittle, W. Ford. "Is Nature Really Motherly?" *CoEvolution Quarterly* 29 (1981): 58–65.

Dubos, René. "Gaia and Creative Evolution." *Nature* 282 (1979): 154–55.

Environment website. "Gaia Hypothesis." Accessed December 23, 2013. http://www.environment.gen.tr/gaia/70-gaia-hypothesis.html.

Flint, James. "Earth—The Final Conflict." The *Daily Telegraph*, February 6, 2006. Accessed December 23, 2013. http://www.telegraph.co.uk/culture/books/3649909/Earth-the-final-conflict.html.

Forbes, Peter. "Jim'll Fix It." The *Guardian*, February 21, 2009. Accessed December 23, 2013. http://www.theguardian.com/culture/2009/feb/21/james-lovelock-gaia-book-review.

Gaia Theory Conference. "Gaia Theory Conference at George Mason University." Arlington County. Accessed December 27, 2013. http://www.gaiatheory.org/2006-conference/.

Goodell, Jeff. "James Lovelock, the Prophet." *Rolling Stone*, November 1, 2007. Accessed December 23, 2013. http://www.rollingstone.com/politics/news/james-lovelock-the-prophet-20071101.

Gould, Stephen J. "Kropotkin Was No Crackpot." *Natural History* 106 (1997): 12–21.

Gray, John. "James Lovelock: A Man for All Seasons." *New Statesman*, March 27, 2013. Accessed December 21, 2013. http://www.newstatesman.com/culture/culture/2013/03/james-lovelock-man-all-seasons.

"The Revenge of Gaia, by James Lovelock." The *Independent*, January 27, 2006. Accessed December 24, 2013. http://www.independent.co.uk/arts-entertainment/books/reviews/the-revenge-of-gaia-by-james-lovelock-6110631.htmlhttp://www.independent.co.uk/arts-entertainment/books/reviews/the-revenge-of-gaia-by-james-lovelock-524635.html.

Hauk, Marna, Judith Landsman, Jeanine M. Canty, and Noël Cox Caniglia. "Gaian Methodologies: An Emergent Confluence of Sustainability Research Innovation." Paper presented at the Association for the Advancement of Sustainability in Higher Education Conference, Denver, October 10–12, 2010.

Irvine, Ian. "James Lovelock: The Green Man." The *Independent*, December 3, 2005. Accessed October 10, 2013. http://www.independent.co.uk/news/people/profiles/james-lovelock-the-green-man-517953.html.

James Lovelock's official website. "Curriculum Vitae." Accessed December 29, 2013. http://www.jameslovelock.org/page2.html.

Kasting, James F. "The Gaia Hypothesis Is Still Giving Us Feedback." *Nautilus* 12 (2014).

Kauffman, Eric G. "The Gaia Controversy: AGU's Chapman Conference." *Eos, Transactions of the American Geophysical Union* 69, no. 31 (1988): 763–64.

Kennedy, John F., Presidential Library and Museum. "Space Program." Accessed January 8, 2016. http://www.jfklibrary.org/JFK/JFK-in-History/Space-Program.aspx.

Kirchner, James W. "Gaia Hypothesis: Can it Be Tested?" *Reviews of Geophysics* 27, no. 2 (1989): 223–35.

"The Gaia Hypothesis: Conjectures and Refutations." *Climatic Change* 58 (2003): 21–45.

"The Gaia Hypothesis: Fact, Theory, and Wishful Thinking." *Climatic* Change 52 (2002): 391–408.

Lenton, Timothy M. "Gaia and Natural Selection." *Nature* 394 (1998): 439–47.

Lenton, Timothy M., and Andrew Watson. *Revolutions that Made the Earth.* Oxford: Oxford University Press, 2013.

Lovelock, James E. *The Ages of Gaia: A Biography of our Living Earth*, Rev. ed. Oxford: Oxford University Press, 1995.

Gaia: A New Look at Life on Earth. Rev. ed. Oxford: Oxford University Press, 2000.

Gaia: Medicine for an Ailing Planet. London: Gaia Books, 2005.

"Geophysiology, the Science of Gaia." *Reviews of Geophysics* 27 (1989): 215–22.

Healing Gaia: The Practical Science of Planetary Medicine. Oxford: Oxford University Press, 2001.

Homage to Gaia: The Life of an Independent Scientist, Rev. ed. London: Souvenir Press Ltd., 2014.

"Gaia: The Living Earth." *Nature* 426 (2003): 769–70.

The Revenge of Gaia: Why the Earth is Fighting Back—and How we Can Still Save Humanity. London: Penguin Books, 2007.

A Rough Ride to the Future. London: Penguin Books, 2015.

The Vanishing Face of Gaia: A Final Warning. London: Penguin Books, 2010.

"Wollaston Medal Citation." Accessed March 4, 2016. http://www.jameslovelock.org/page7.html.

Lovelock, James E., and Sidney Epton. "The Quest for Gaia." *New Scientist* 65, no. 935 (1975): 304–09.

Lovelock, James E., and C. E. Giffin. "Planetary Atmospheres: Compositional and Other Changes Associated with the Presence of Life." *Advances in the Astronautical Sciences* 25 (1969): 179–93.

Lovelock, James E., and Lynn Margulis. "Atmospheric Homeostasis by and for the Biosphere: The Gaia Hypothesis." *Tellus* 26, nos. 1–2 (1974): 2–10.

McKie, Robin. "Gaia's Warrior." *Green Lifestyle Magazine*, July/August 2007.

Mellanby, Kenneth. "Living with the Earth Mother." *New Scientist* 84 (1979): 41.

Midgley, Mary. "Great Thinkers—James Lovelock." *New Statesman*, 14 July 2003.

Ogle, Martin. "The Gaia Theory: Scientific Model and Metaphor for the 21st Century." *Revista Umbral (Threshold Magazine)* 1 (2009): 99–106.

Ravilious, Kate. "Perfect Harmony." The *Guardian*, April 28, 2008. Accessed December 30, 2013. www.theguardian.com/science/2008/apr/28/scienceofclimatechange.biodiversity.

Schneider, Stephen H., and Penelope J. Boston, eds. *Scientists on Gaia*. Cambridge: The MIT Press, 1993.

Schneider, Stephen H., James R. Miller, Eileen Crist, and Penelope J. Boston, eds. *Scientists Debate Gaia: The Next Century*. Cambridge: The MIT Press, 2004.

Tickell, Crispin. "Scientists on Gaia." The *Financial Times* 2002. Accessed December 23, 2013. http://www.crispintickell.com/page19.html.

Turney, Jon. *Lovelock and Gaia: Signs of Life*. New York: Columbia University Press, 2003.

Tyrrell, Toby. *On Gaia: A Critical Investigation of the Relationship Between Life and Earth*. Princeton: Princeton University Press, 2013.

Velasquez-Manoff, Moises. "The Medea Hypothesis: A Response to the Gaia Hypothesis." *The Christian Science Monitor*, February 12, 2010. Accessed December 28, 2013. http://www.csmonitor.com/Environment/Bright-Green/2010/0212/The-Medea-Hypothesis-A-response-to-the-Gaia-hypothesis.

Wallace, Richard R., and Bryan G. Norton. "Policy Implications of Gaian Theory." *Ecological Economics* 6 (1992): 103–18.

Ward, Peter. *The Medea Hypothesis: Is Life on Earth Ultimately Self-Destructive?* Princeton: Princeton University Press, 2009.

Watson, Andrew J., and James E. Lovelock. "Biological Homeostasis of the Global Environment: The Parable of Daisyworld." *Tellus* 35B (1983): 286–89.

THE MACAT LIBRARY
BY DISCIPLINE

AFRICANA STUDIES

Chinua Achebe's *An Image of Africa: Racism in Conrad's Heart of Darkness*
W. E. B. Du Bois's *The Souls of Black Folk*
Zora Neale Huston's *Characteristics of Negro Expression*
Martin Luther King Jr's *Why We Can't Wait*
Toni Morrison's *Playing in the Dark: Whiteness in the American Literary Imagination*

ANTHROPOLOGY

Arjun Appadurai's *Modernity at Large: Cultural Dimensions of Globalisation*
Philippe Ariès's *Centuries of Childhood*
Franz Boas's *Race, Language and Culture*
Kim Chan & Renée Mauborgne's *Blue Ocean Strategy*
Jared Diamond's *Guns, Germs & Steel: the Fate of Human Societies*
Jared Diamond's *Collapse: How Societies Choose to Fail or Survive*
E. E. Evans-Pritchard's *Witchcraft, Oracles and Magic Among the Azande*
James Ferguson's *The Anti-Politics Machine*
Clifford Geertz's *The Interpretation of Cultures*
David Graeber's *Debt: the First 5000 Years*
Karen Ho's *Liquidated: An Ethnography of Wall Street*
Geert Hofstede's *Culture's Consequences: Comparing Values, Behaviors, Institutes and Organizations across Nations*
Claude Lévi-Strauss's *Structural Anthropology*
Jay Macleod's *Ain't No Makin' It: Aspirations and Attainment in a Low-Income Neighborhood*
Saba Mahmood's *The Politics of Piety: The Islamic Revival and the Feminist Subjec*t
Marcel Mauss's *The Gift*

BUSINESS

Jean Lave & Etienne Wenger's *Situated Learning*
Theodore Levitt's *Marketing Myopia*
Burton G. Malkiel's *A Random Walk Down Wall Street*
Douglas McGregor's *The Human Side of Enterprise*
Michael Porter's *Competitive Strategy: Creating and Sustaining Superior Performance*
John Kotter's *Leading Change*
C. K. Prahalad & Gary Hamel's *The Core Competence of the Corporation*

CRIMINOLOGY

Michelle Alexander's *The New Jim Crow: Mass Incarceration in the Age of Colorblindness*
Michael R. Gottfredson & Travis Hirschi's *A General Theory of Crime*
Richard Herrnstein & Charles A. Murray's *The Bell Curve: Intelligence and Class Structure in American Life*
Elizabeth Loftus's *Eyewitness Testimony*
Jay Macleod's *Ain't No Makin' It: Aspirations and Attainment in a Low-Income Neighborhood*
Philip Zimbardo's *The Lucifer Effect*

ECONOMICS

Janet Abu-Lughod's *Before European Hegemony*
Ha-Joon Chang's *Kicking Away the Ladder*
David Brion Davis's *The Problem of Slavery in the Age of Revolution*
Milton Friedman's *The Role of Monetary Policy*
Milton Friedman's *Capitalism and Freedom*
David Graeber's *Debt: the First 5000 Years*
Friedrich Hayek's *The Road to Serfdom*
Karen Ho's *Liquidated: An Ethnography of Wall Street*

John Maynard Keynes's *The General Theory of Employment, Interest and Money*
Charles P. Kindleberger's *Manias, Panics and Crashes*
Robert Lucas's *Why Doesn't Capital Flow from Rich to Poor Countries?*
Burton G. Malkiel's *A Random Walk Down Wall Street*
Thomas Robert Malthus's *An Essay on the Principle of Population*
Karl Marx's *Capital*
Thomas Piketty's *Capital in the Twenty-First Century*
Amartya Sen's *Development as Freedom*
Adam Smith's *The Wealth of Nations*
Nassim Nicholas Taleb's *The Black Swan: The Impact of the Highly Improbable*
Amos Tversky's & Daniel Kahneman's *Judgment under Uncertainty: Heuristics and Biases*
Mahbub Ul Haq's *Reflections on Human Development*
Max Weber's *The Protestant Ethic and the Spirit of Capitalism*

FEMINISM AND GENDER STUDIES

Judith Butler's *Gender Trouble*
Simone De Beauvoir's *The Second Sex*
Michel Foucault's *History of Sexuality*
Betty Friedan's *The Feminine Mystique*
Saba Mahmood's *The Politics of Piety: The Islamic Revival and the Feminist Subject*
Joan Wallach Scott's *Gender and the Politics of History*
Mary Wollstonecraft's *A Vindication of the Rights of Woman*
Virginia Woolf's *A Room of One's Own*

GEOGRAPHY

The Brundtland Report's *Our Common Future*
Rachel Carson's *Silent Spring*
Charles Darwin's *On the Origin of Species*
James Ferguson's *The Anti-Politics Machine*
Jane Jacobs's *The Death and Life of Great American Cities*
James Lovelock's *Gaia: A New Look at Life on Earth*
Amartya Sen's *Development as Freedom*
Mathis Wackernagel & William Rees's *Our Ecological Footprint*

HISTORY

Janet Abu-Lughod's *Before European Hegemony*
Benedict Anderson's *Imagined Communities*
Bernard Bailyn's *The Ideological Origins of the American Revolution*
Hanna Batatu's *The Old Social Classes And The Revolutionary Movements Of Iraq*
Christopher Browning's *Ordinary Men: Reserve Police Batallion 101 and the Final Solution in Poland*
Edmund Burke's *Reflections on the Revolution in France*
William Cronon's *Nature's Metropolis: Chicago And The Great West*
Alfred W. Crosby's *The Columbian Exchange*
Hamid Dabashi's *Iran: A People Interrupted*
David Brion Davis's *The Problem of Slavery in the Age of Revolution*
Nathalie Zemon Davis's *The Return of Martin Guerre*
Jared Diamond's *Guns, Germs & Steel: the Fate of Human Societies*
Frank Dikotter's *Mao's Great Famine*
John W Dower's *War Without Mercy: Race And Power In The Pacific War*
W. E. B. Du Bois's *The Souls of Black Folk*
Richard J. Evans's *In Defence of History*
Lucien Febvre's *The Problem of Unbelief in the 16th Century*
Sheila Fitzpatrick's *Everyday Stalinism*

The Macat Library By Discipline

Eric Foner's *Reconstruction: America's Unfinished Revolution, 1863-1877*
Michel Foucault's *Discipline and Punish*
Michel Foucault's *History of Sexuality*
Francis Fukuyama's *The End of History and the Last Man*
John Lewis Gaddis's *We Now Know: Rethinking Cold War History*
Ernest Gellner's *Nations and Nationalism*
Eugene Genovese's *Roll, Jordan, Roll: The World the Slaves Made*
Carlo Ginzburg's *The Night Battles*
Daniel Goldhagen's *Hitler's Willing Executioners*
Jack Goldstone's *Revolution and Rebellion in the Early Modern World*
Antonio Gramsci's *The Prison Notebooks*
Alexander Hamilton, John Jay & James Madison's *The Federalist Papers*
Christopher Hill's *The World Turned Upside Down*
Carole Hillenbrand's *The Crusades: Islamic Perspectives*
Thomas Hobbes's *Leviathan*
Eric Hobsbawm's *The Age Of Revolution*
John A. Hobson's *Imperialism: A Study*
Albert Hourani's *History of the Arab Peoples*
Samuel P. Huntington's *The Clash of Civilizations and the Remaking of World Order*
C. L. R. James's *The Black Jacobins*
Tony Judt's *Postwar: A History of Europe Since 1945*
Ernst Kantorowicz's *The King's Two Bodies: A Study in Medieval Political Theology*
Paul Kennedy's *The Rise and Fall of the Great Powers*
Ian Kershaw's *The "Hitler Myth": Image and Reality in the Third Reich*
John Maynard Keynes's *The General Theory of Employment, Interest and Money*
Charles P. Kindleberger's *Manias, Panics and Crashes*
Martin Luther King Jr's *Why We Can't Wait*
Henry Kissinger's *World Order: Reflections on the Character of Nations and the Course of History*
Thomas Kuhn's *The Structure of Scientific Revolutions*
Georges Lefebvre's *The Coming of the French Revolution*
John Locke's *Two Treatises of Government*
Niccolò Machiavelli's *The Prince*
Thomas Robert Malthus's *An Essay on the Principle of Population*
Mahmood Mamdani's *Citizen and Subject: Contemporary Africa And The Legacy Of Late Colonialism*
Karl Marx's *Capital*
Stanley Milgram's *Obedience to Authority*
John Stuart Mill's *On Liberty*
Thomas Paine's *Common Sense*
Thomas Paine's *Rights of Man*
Geoffrey Parker's *Global Crisis: War, Climate Change and Catastrophe in the Seventeenth Century*
Jonathan Riley-Smith's *The First Crusade and the Idea of Crusading*
Jean-Jacques Rousseau's *The Social Contract*
Joan Wallach Scott's *Gender and the Politics of History*
Theda Skocpol's *States and Social Revolutions*
Adam Smith's *The Wealth of Nations*
Timothy Snyder's *Bloodlands: Europe Between Hitler and Stalin*
Sun Tzu's *The Art of War*
Keith Thomas's *Religion and the Decline of Magic*
Thucydides's *The History of the Peloponnesian War*
Frederick Jackson Turner's *The Significance of the Frontier in American History*
Odd Arne Westad's *The Global Cold War: Third World Interventions And The Making Of Our Times*

LITERATURE

Chinua Achebe's *An Image of Africa: Racism in Conrad's Heart of Darkness*
Roland Barthes's *Mythologies*
Homi K. Bhabha's *The Location of Culture*
Judith Butler's *Gender Trouble*
Simone De Beauvoir's *The Second Sex*
Ferdinand De Saussure's *Course in General Linguistics*
T. S. Eliot's *The Sacred Wood: Essays on Poetry and Criticism*
Zora Neale Huston's *Characteristics of Negro Expression*
Toni Morrison's *Playing in the Dark: Whiteness in the American Literary Imagination*
Edward Said's *Orientalism*
Gayatri Chakravorty Spivak's *Can the Subaltern Speak?*
Mary Wollstonecraft's *A Vindication of the Rights of Women*
Virginia Woolf's *A Room of One's Own*

PHILOSOPHY

Elizabeth Anscombe's *Modern Moral Philosophy*
Hannah Arendt's *The Human Condition*
Aristotle's *Metaphysics*
Aristotle's *Nicomachean Ethics*
Edmund Gettier's *Is Justified True Belief Knowledge?*
Georg Wilhelm Friedrich Hegel's *Phenomenology of Spirit*
David Hume's *Dialogues Concerning Natural Religion*
David Hume's *The Enquiry for Human Understanding*
Immanuel Kant's *Religion within the Boundaries of Mere Reason*
Immanuel Kant's *Critique of Pure Reason*
Søren Kierkegaard's *The Sickness Unto Death*
Søren Kierkegaard's *Fear and Trembling*
C. S. Lewis's *The Abolition of Man*
Alasdair MacIntyre's *After Virtue*
Marcus Aurelius's *Meditations*
Friedrich Nietzsche's *On the Genealogy of Morality*
Friedrich Nietzsche's *Beyond Good and Evil*
Plato's *Republic*
Plato's *Symposium*
Jean-Jacques Rousseau's *The Social Contract*
Gilbert Ryle's *The Concept of Mind*
Baruch Spinoza's *Ethics*
Sun Tzu's *The Art of War*
Ludwig Wittgenstein's *Philosophical Investigations*

POLITICS

Benedict Anderson's *Imagined Communities*
Aristotle's *Politics*
Bernard Bailyn's *The Ideological Origins of the American Revolution*
Edmund Burke's *Reflections on the Revolution in France*
John C. Calhoun's *A Disquisition on Government*
Ha-Joon Chang's *Kicking Away the Ladder*
Hamid Dabashi's *Iran: A People Interrupted*
Hamid Dabashi's *Theology of Discontent: The Ideological Foundation of the Islamic Revolution in Iran*
Robert Dahl's *Democracy and its Critics*
Robert Dahl's *Who Governs?*
David Brion Davis's *The Problem of Slavery in the Age of Revolution*

The Macat Library By Discipline

Alexis De Tocqueville's *Democracy in America*
James Ferguson's *The Anti-Politics Machine*
Frank Dikotter's *Mao's Great Famine*
Sheila Fitzpatrick's *Everyday Stalinism*
Eric Foner's *Reconstruction: America's Unfinished Revolution, 1863-1877*
Milton Friedman's *Capitalism and Freedom*
Francis Fukuyama's *The End of History and the Last Man*
John Lewis Gaddis's *We Now Know: Rethinking Cold War History*
Ernest Gellner's *Nations and Nationalism*
David Graeber's *Debt: the First 5000 Years*
Antonio Gramsci's *The Prison Notebooks*
Alexander Hamilton, John Jay & James Madison's *The Federalist Papers*
Friedrich Hayek's *The Road to Serfdom*
Christopher Hill's *The World Turned Upside Down*
Thomas Hobbes's *Leviathan*
John A. Hobson's *Imperialism: A Study*
Samuel P. Huntington's *The Clash of Civilizations and the Remaking of World Order*
Tony Judt's *Postwar: A History of Europe Since 1945*
David C. Kang's *China Rising: Peace, Power and Order in East Asia*
Paul Kennedy's *The Rise and Fall of Great Powers*
Robert Keohane's *After Hegemony*
Martin Luther King Jr.'s *Why We Can't Wait*
Henry Kissinger's *World Order: Reflections on the Character of Nations and the Course of History*
John Locke's *Two Treatises of Government*
Niccolò Machiavelli's *The Prince*
Thomas Robert Malthus's *An Essay on the Principle of Population*
Mahmood Mamdani's *Citizen and Subject: Contemporary Africa And The Legacy Of Late Colonialism*
Karl Marx's *Capital*
John Stuart Mill's *On Liberty*
John Stuart Mill's *Utilitarianism*
Hans Morgenthau's *Politics Among Nations*
Thomas Paine's *Common Sense*
Thomas Paine's *Rights of Man*
Thomas Piketty's *Capital in the Twenty-First Century*
Robert D. Putman's *Bowling Alone*
John Rawls's *Theory of Justice*
Jean-Jacques Rousseau's *The Social Contract*
Theda Skocpol's *States and Social Revolutions*
Adam Smith's *The Wealth of Nations*
Sun Tzu's *The Art of War*
Henry David Thoreau's *Civil Disobedience*
Thucydides's *The History of the Peloponnesian War*
Kenneth Waltz's *Theory of International Politics*
Max Weber's *Politics as a Vocation*
Odd Arne Westad's *The Global Cold War: Third World Interventions And The Making Of Our Times*

POSTCOLONIAL STUDIES

Roland Barthes's *Mythologies*
Frantz Fanon's *Black Skin, White Masks*
Homi K. Bhabha's *The Location of Culture*
Gustavo Gutiérrez's *A Theology of Liberation*
Edward Said's *Orientalism*
Gayatri Chakravorty Spivak's *Can the Subaltern Speak?*

PSYCHOLOGY

Gordon Allport's *The Nature of Prejudice*
Alan Baddeley & Graham Hitch's *Aggression: A Social Learning Analysis*
Albert Bandura's *Aggression: A Social Learning Analysis*
Leon Festinger's *A Theory of Cognitive Dissonance*
Sigmund Freud's *The Interpretation of Dreams*
Betty Friedan's *The Feminine Mystique*
Michael R. Gottfredson & Travis Hirschi's *A General Theory of Crime*
Eric Hoffer's *The True Believer: Thoughts on the Nature of Mass Movements*
William James's *Principles of Psychology*
Elizabeth Loftus's *Eyewitness Testimony*
A. H. Maslow's *A Theory of Human Motivation*
Stanley Milgram's *Obedience to Authority*
Steven Pinker's *The Better Angels of Our Nature*
Oliver Sacks's *The Man Who Mistook His Wife For a Hat*
Richard Thaler & Cass Sunstein's *Nudge: Improving Decisions About Health, Wealth and Happiness*
Amos Tversky's *Judgment under Uncertainty: Heuristics and Biases*
Philip Zimbardo's *The Lucifer Effect*

SCIENCE

Rachel Carson's *Silent Spring*
William Cronon's *Nature's Metropolis: Chicago And The Great West*
Alfred W. Crosby's *The Columbian Exchange*
Charles Darwin's *On the Origin of Species*
Richard Dawkin's *The Selfish Gene*
Thomas Kuhn's *The Structure of Scientific Revolutions*
Geoffrey Parker's *Global Crisis: War, Climate Change and Catastrophe in the Seventeenth Century*
Mathis Wackernagel & William Rees's *Our Ecological Footprint*

SOCIOLOGY

Michelle Alexander's *The New Jim Crow: Mass Incarceration in the Age of Colorblindness*
Gordon Allport's *The Nature of Prejudice*
Albert Bandura's *Aggression: A Social Learning Analysis*
Hanna Batatu's *The Old Social Classes And The Revolutionary Movements Of Iraq*
Ha-Joon Chang's *Kicking Away the Ladder*
W. E. B. Du Bois's *The Souls of Black Folk*
Émile Durkheim's *On Suicide*
Frantz Fanon's *Black Skin, White Masks*
Frantz Fanon's *The Wretched of the Earth*
Eric Foner's *Reconstruction: America's Unfinished Revolution, 1863-1877*
Eugene Genovese's *Roll, Jordan, Roll: The World the Slaves Made*
Jack Goldstone's *Revolution and Rebellion in the Early Modern World*
Antonio Gramsci's *The Prison Notebooks*
Richard Herrnstein & Charles A Murray's *The Bell Curve: Intelligence and Class Structure in American Life*
Eric Hoffer's *The True Believer: Thoughts on the Nature of Mass Movements*
Jane Jacobs's *The Death and Life of Great American Cities*
Robert Lucas's *Why Doesn't Capital Flow from Rich to Poor Countries?*
Jay Macleod's *Ain't No Makin' It: Aspirations and Attainment in a Low Income Neighborhood*
Elaine May's *Homeward Bound: American Families in the Cold War Era*
Douglas McGregor's *The Human Side of Enterprise*
C. Wright Mills's *The Sociological Imagination*

The Macat Library By Discipline

Thomas Piketty's *Capital in the Twenty-First Century*
Robert D. Putman's *Bowling Alone*
David Riesman's *The Lonely Crowd: A Study of the Changing American Character*
Edward Said's *Orientalism*
Joan Wallach Scott's *Gender and the Politics of History*
Theda Skocpol's *States and Social Revolutions*
Max Weber's *The Protestant Ethic and the Spirit of Capitalism*

THEOLOGY

Augustine's *Confessions*
Benedict's *Rule of St Benedict*
Gustavo Gutiérrez's *A Theology of Liberation*
Carole Hillenbrand's *The Crusades: Islamic Perspectives*
David Hume's *Dialogues Concerning Natural Religion*
Immanuel Kant's *Religion within the Boundaries of Mere Reason*
Ernst Kantorowicz's *The King's Two Bodies: A Study in Medieval Political Theology*
Søren Kierkegaard's *The Sickness Unto Death*
C. S. Lewis's *The Abolition of Man*
Saba Mahmood's *The Politics of Piety: The Islamic Revival and the Feminist Subject*
Baruch Spinoza's *Ethics*
Keith Thomas's *Religion and the Decline of Magic*

COMING SOON

Chris Argyris's *The Individual and the Organisation*
Seyla Benhabib's *The Rights of Others*
Walter Benjamin's *The Work Of Art in the Age of Mechanical Reproduction*
John Berger's *Ways of Seeing*
Pierre Bourdieu's *Outline of a Theory of Practice*
Mary Douglas's *Purity and Danger*
Roland Dworkin's *Taking Rights Seriously*
James G. March's *Exploration and Exploitation in Organisational Learning*
Ikujiro Nonaka's *A Dynamic Theory of Organizational Knowledge Creation*
Griselda Pollock's *Vision and Difference*
Amartya Sen's *Inequality Re-Examined*
Susan Sontag's *On Photography*
Yasser Tabbaa's *The Transformation of Islamic Art*
Ludwig von Mises's *Theory of Money and Credit*

Macat Pairs

*Analyse historical and modern issues
from opposite sides of an argument.
Pairs include:*

INTERNATIONAL RELATIONS IN THE 21ST CENTURY

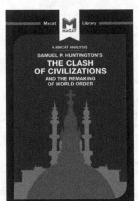

Samuel P. Huntington's
The Clash of Civilisations

In his highly influential 1996 book, Huntington offers a vision of a post-Cold War world in which conflict takes place not between competing ideologies but between cultures. The worst clash, he argues, will be between the Islamic world and the West: the West's arrogance and belief that its culture is a "gift" to the world will come into conflict with Islam's obstinacy and concern that its culture is under attack from a morally decadent "other."

Clash inspired much debate between different political schools of thought. But its greatest impact came in helping define American foreign policy in the wake of the 2001 terrorist attacks in New York and Washington.

Francis Fukuyama's
The End of History and the Last Man

Published in 1992, *The End of History and the Last Man* argues that capitalist democracy is the final destination for all societies. Fukuyama believed democracy triumphed during the Cold War because it lacks the "fundamental contradictions" inherent in communism and satisfies our yearning for freedom and equality. Democracy therefore marks the endpoint in the evolution of ideology, and so the "end of history." There will still be "events," but no fundamental change in ideology.

Macat Pairs

Analyse historical and modern issues from opposite sides of an argument.
Pairs include:

ARE WE FUNDAMENTALLY GOOD - OR BAD?

Steven Pinker's
The Better Angels of Our Nature

Stephen Pinker's gloriously optimistic 2011 book argues that, despite humanity's biological tendency toward violence, we are, in fact, less violent today than ever before. To prove his case, Pinker lays out pages of detailed statistical evidence. For him, much of the credit for the decline goes to the eighteenth-century Enlightenment movement, whose ideas of liberty, tolerance, and respect for the value of human life filtered down through society and affected how people thought. That psychological change led to behavioral change—and overall we became more peaceful. Critics countered that humanity could never overcome the biological urge toward violence; others argued that Pinker's statistics were flawed.

Philip Zimbardo's
The Lucifer Effect

Some psychologists believe those who commit cruelty are innately evil. Zimbardo disagrees. In *The Lucifer Effect*, he argues that sometimes good people do evil things simply because of the situations they find themselves in, citing many historical examples to illustrate his point. Zimbardo details his 1971 Stanford prison experiment, where ordinary volunteers playing guards in a mock prison rapidly became abusive. But he also describes the tortures committed by US army personnel in Iraq's Abu Ghraib prison in 2003—and how he himself testified in defence of one of those guards. committed by US army personnel in Iraq's Abu Ghraib prison in 2003—and how he himself testified in defence of one of those guards.

Macat Pairs

Analyse historical and modern issues from opposite sides of an argument. Pairs include:

HOW WE RELATE TO EACH OTHER AND SOCIETY

Jean-Jacques Rousseau's
The Social Contract

Rousseau's famous work sets out the radical concept of the 'social contract': a give-and-take relationship between individual freedom and social order.

If people are free to do as they like, governed only by their own sense of justice, they are also vulnerable to chaos and violence. To avoid this, Rousseau proposes, they should agree to give up some freedom to benefit from the protection of social and political organization. But this deal is only just if societies are led by the collective needs and desires of the people, and able to control the private interests of individuals. For Rousseau, the only legitimate form of government is rule by the people.

Robert D. Putnam's
Bowling Alone

In *Bowling Alone*, Robert Putnam argues that Americans have become disconnected from one another and from the institutions of their common life, and investigates the consequences of this change.

Looking at a range of indicators, from membership in formal organizations to the number of invitations being extended to informal dinner parties, Putnam demonstrates that Americans are interacting less and creating less "social capital" – with potentially disastrous implications for their society.

It would be difficult to overstate the impact of *Bowling Alone*, one of the most frequently cited social science publications of the last half-century.

Macat analyses are available from all good bookshops and libraries.

Access hundreds of analyses through one, multimedia tool.
Join free for one month **library.macat.com**

Printed in the United States
by Baker & Taylor Publisher Services